INFINITE
ENERGY
TECHNOLOGIES

"With humanity facing the dismaying prospects of global ecological collapse and geopolitical chaos, there is an urgent need for clear solutions-based guidance that penetrates our dulled consciousness and pulls us back from the precipice. *Infinite Energy Technologies* delivers such guidance. Through a powerfully resonant combination of new energy science, societal analysis, and spiritual insight, Finley Eversole's compilation shakes us awake from our dangerous stupor. The wise voices in this anthology make a compelling case for the immediate embrace of a new wave of energy technologies that is key to launching an era of shared abundance, planetary healing, and unprecedented creativity. I pray that millions will heed this call for action without delay and lead the transformation so desperately needed on our imperiled planet."

JOEL GARBON, PRESIDENT OF NEW ENERGY MOVEMENT
AND COAUTHOR OF *BREAKTHROUGH POWER*

INFINITE ENERGY TECHNOLOGIES

■▪■

Dedicated to geniuses of vision:

John Worrell Keely

Nikola Tesla

Viktor Schauberger

Georges Lakhovsky

Royal Raymond Rife

Thomas Townsend Brown

Martin Fleischmann and Stanley Pons

And in memory of

Eugene Franklin Mallove

Founder of Infinite Energy *magazine*

and

Brian O'Leary

Founder of the New Energy Movement

The writing is on the wall. The fossil fuel age is about to end.

EUGENE MALLOVE

Take courage. The human race is divine.

PYTHAGORAS

Humanity will become eventually the planetary savior.

DJWHAL KHUL

The people of Earth are a superpower themselves, if united.

JEANE MANNING

This time, like all times, is a very good one, if we but know what to do with it.

RALPH WALDO EMERSON

We have it in our power to begin the world over again.

THOMAS PAINE

We stand at the gateway of the new world, of the new age and its new civilizations, ideals and culture. What is coming is a civilization of a different yet still material nature, but animated by a growing registration by the masses everywhere of an emerging spiritual objective which will transform all life and give new value and purpose to that which is material.

DJWHAL KHUL

Another world is not only possible, she is on her way. On a quiet day, I can hear her breathing.

ARUNDHATI ROY

There is always a new world in process of forming. . . . The best is yet to be.

DJWHAL KHUL

Contents

Acknowledgments — xii

Foreword — xiii

John L. Petersen

Introduction — 1

Finley Eversole, Ph.D.

PART 1

Back to the Future: The Legacy of the Visionaries

1 Nikola Tesla: Electrical Savant — 13

Marc J. Seifer, Ph.D.

2 John Worrell Keely: Free-Energy Pioneer— a New Chapter — 43

Theo Paijmans

3 Viktor Schauberger: A Brief Overview of His Theories on Energy, Motion, and Water — 74

Callum Coats

4 ROYAL RAYMOND RIFE: THE FATE OF
   COMPASSION AND THE CANCER CURE
   THAT WORKED   120

   Gerry Vassilatos

5 T. TOWNSEND BROWN: THE SUPPRESSION
   OF ANTIGRAVITY TECHNOLOGY   156

   Jeane Manning

PART 2

Infinite Energy:
A New Science for a Pollution-Free World

6 THE SUSTAINABLE TECHNOLOGY SOLUTION
   REVOLUTION: A UNIVERSAL APPEAL   171

   Brian O'Leary, Ph.D.

7 POWER FOR THE PEOPLE—FROM WATER   186

   Jeane Manning

8 IMAGINE A FREE-ENERGY FUTURE
   FOR ALL OF HUMANITY   210

   Steven M. Greer, M.D.

9 ENERGY TECHNOLOGIES FOR
   THE TWENTY-FIRST CENTURY   223

   Theodore C. Loder III, Ph.D.

10 HARNESSING NATURE'S FREE ENERGY:
   THE SEARL EFFECT GENERATOR   240

   John R. R. Searl and John A. Thomas Jr.

11 COLD FUSION: THE END TO
   CONVENTIONAL ENERGY AND THE
   START OF SOCIAL REORGANIZATION   254

   Edmund Storms, Ph.D.

**12   Zero Point Energy Can Power
the Future**   276

Thomas Valone, Ph.D., P.E.

Afterword   302

Brian O'Leary, Ph.D.

◼▪◼

**APPENDIX A   The Four Occupations of
Planet Earth**   306

Tom Engelhardt

APPENDIX B   Evidence of Cosmic Community   315

Finley Eversole, Ph.D.

APPENDIX C   The Earth Charter   325

▪◼▪

Notes   336

Additional Resources   353

Contributors   365

Index   371

ACKNOWLEDGMENTS

My first thanks go to all of the authors who have given generously of their time and wisdom to write the chapters contained in this volume and for their quick responses to my many communications and requests. Jeane Manning, Ed Storms, and Emily Greer—wife of Steven Greer—frequently served as consults when I had questions. I also want to thank Dale Pond for providing some of the photos for the chapter on John Worrell Keely.

Thanks also to futurist John Petersen of the Arlington Institute for writing the foreword.

I would especially like to thank my publisher Ehud Sperling, and Jon Graham, acquisitions editor, for approving this project when it was still only a one-paragraph idea and for patiently waiting for it to become a reality.

Last, but not least, my thanks also go the Inner Traditions staff who have worked so hard to bring this book forth—Jeanie Levitan, managing editor, Jon Desautels for the cover design, and above all to my project editor, Anne Dillon, whose uncommon dedication and meticulous attention to every detail of this project have helped make this book be the best it can be.

Infinite Energy Technologies has been a collaborative project in every sense of the word. May it move the public to think of *energy solutions* in far broader terms than those which now dominate the public agenda.

FOREWORD

John L. Petersen

We are living in unprecedented times, but, of course, everyone has said that at any given period in the past. Nevertheless, technically it's true. Every year is a fresh, new one that might seem familiar, but essentially is not. Unless all change could be eliminated, we're necessarily producing new realities at every moment that have never existed before.

Parallels with historical times, at best, therefore, reflect only a very rough congruity with an earlier time that certainly did not have the technology, communications, ideas, and values of the present. So, sure, these are unprecedented times.

But in important ways, this time it is *really* unprecedented. There is always change, but the rate of change that we are experiencing these days has never been seen before—and it is accelerating exponentially. That means that if present trends continue, every week or month or year going forward will produce *significantly* more change than the previous one. Humans have never experienced this rate of change before.

Let me give you an example. Futurist Ray Kurzweil, in his important book *The Singularity Is Near,* cataloged the rate of technological change in many different dimensions. His bottom-line assessment was that our present century will see one thousand times the technological change of the past century—during which the automobile, airplane,

Internet, and nuclear wars emerged. Transportation rates went from that limited by the gallop of a horse to chemically propelled spacecraft that traverse more than fifteen thousand miles in an hour. And, of course, we visited the moon.

Now, think about what one thousand times that change would be. What kind of a world might show up in one hundred years if we live through one thousand times the change of the twentieth century? Well, you can't reasonably imagine it. No one can. The implications are so great that you are immediately driven into science fiction land, where all of the current "experts" just dismiss you with a wave of a hand.

Try it. With two compounded orders of magnitude change over the period of a century, you could literally find yourself in a place where humans didn't eat food or drink water (which would eliminate agriculture). They might be able to read minds telepathically and be able to visually read the energetic fields of anyone they looked at—immediately knowing about the person's past experiences, present feelings, and honesty of statements. Just that, of course, would eliminate all politicians and advertising!

But maybe, as some sources seriously suggest, you could manifest physical things at will—just by focusing your mind. Think of what that would do to the notion of economics as we know it. In this handful of future human characteristics, you'd also be able to transport yourself wherever you wanted by thinking yourself there. In that world, no one would know what airplanes were.

You might think that what I've just described is far-fetched, and if so, then you just made my point. Even though there are credible analysts and observers who seriously propose that the above changes will happen in far less time than a century, change of this sort is more than we can reasonably understand and visualize. Just to parse it down to the next decade—seventy to eighty times the change of the past century—boggles the mind!

Well, it's my business to think about these things, and even I have a hard time visualizing how this all might turn out, just because it is so severe and disruptive, but I can tell you a bit about what a revolution of this magnitude means.

First of all, it means that we are in a transition to a new world—a new paradigm. All of this change has direction, and it is leading us to a new world that operates in very different ways.

Second, in this kind of shift, things change fundamentally. We're not talking about adjustments around the edge. The only way to support and sustain this rate of change is if there are extraordinary breakthroughs across almost every sector of human activity.

Already, for example, there are serious efforts afoot to make it possible to control many processes with only your thoughts, and the ability to make physical things invisible has made great strides. In a very short time it will be possible to capture, store, and search through everything you say in any public (or even private) environment and extract from it at will. As this book suggests, unlimited energy and the control of gravity are both in the works.

Third, the tempo accelerates—things change more quickly. The rate of change is increasing, so bigger things are coming faster. And as they converge, these extraordinary events and driving forces interact and cause chain reactions, generating unanticipated consequences. There's a pretty good chance that the inventors of Facebook and Twitter didn't think they were going to be part of bringing down governments, and it's certainly clear that most governments didn't anticipate that this new technology might threaten their ability to govern.

Fourth, much of the change will therefore be strange and unfamiliar. When very rapid, profound, interconnected forces are all in play at the same time, the unanticipated consequences are likely to move quite quickly into threatening the historical and conventional understanding of how things work. Our situation is exacerbated by the fact that significant cosmic changes are influencing the behavior of the sun and therefore major systems (like the climate) on our planet. These are contextual reorganizations that are so large and unprecedented that the underlying systems—agriculture, economy, government, and such—will not be able to respond effectively.

Because of that, human systems will have a hard time adapting to the change. Research has shown that civil and social systems (legal system, education, government, families, etc.) reconfigure themselves thousands of times slower than the rate of technological change that we are experiencing.

Therefore, it is inevitable that the old systems will collapse. They will not have the capability to change fast enough, and in some cases (like the global financial system) have structurally run out of the ability to sustain the status quo.

So, lastly, a new paradigm will emerge from all of this upheaval that only seems chaotic because we're in the middle of it. Something new will arise to fill the vacuum left by the implosion of the legacy systems. If history gives us any indicator of what the new world will be, it is certain that it will be *radically* different from the world that we all now find familiar.

So, what does that all mean relative to this book?

In physical terms, there is no more fundamental and basic influence on the way we live and behave than the availability and form of energy that we use. Every aspect of our lives, food, clothing, shelter, and transportation—and therefore every derivative activity (work, government, recreation, etc.)—changes when the affordable source of energy changes. The modern world has been directly enabled by the discovery, development, and availability of petroleum, for example. When that era ends, many other ways of doing things will also necessarily end.

Thomas Kuhn, influential philosopher of science, famously stated that new paradigms in science emerge only when the leadership of the old generation dies, leaving space for the emergence of the new ideas. What he was saying is that the incumbent system fights new ideas—regardless of whether it is a system of science, education, or spirituality. In all cases, the current generation has vested interests (reputations, income streams, influence, etc.) in the present way things operate. These individuals and organizations have devoted a great deal of time and wealth to building and shaping the present paradigm and would lose a great deal if their ideas, processes, investments, and infrastructure were suddenly deemed obsolete. Like white corpuscles rushing to attack invading germs that are advancing through a break in the integrity of the skin, those with reputations and resources immediately respond to threatening new ideas that could potentially upset what they have worked so hard to put in place.

But Kuhn was describing the dynamics of evolutionary change—change from within. What we're experiencing is revolutionary change that is driven as much by uncontrollable externalities as by internal system dynamics. If rapid climate change sweeps away the assumptions of the past, everyone will have to rethink how things are done. If we begin running out of oil, everyone will be in the business of finding new energy sources. And similarly, if the financial system collapses of its own weight, space will quickly be made for new ideas, and these are just conventional scenarios. If alien life comes by to introduce itself or solar cosmic rays turn on strings of dormant human DNA, suddenly providing us with radical new capabilities—well, all bets are off and new ideas will really prevail!

That's what I think is happening. We are fully into the most significant global revolution in the history of our species. We are about to watch our world turn sideways as the result of the collision of both conventional and unconventional forces, and one big story in that shift will be energy. That's why this book is important.

In many cases, the new ideas that rise to fill the void produced by large-scale change have their origins long before the environment finally presents the opening that allows their proliferation. (Interesting new ideas about alternative financial systems and economies, for example, are now being formed in anticipation of the collapse of the present financial system). More than likely, the system fights those insurgent concepts when they first show up, finding them threatening. Nevertheless, over the years some small groups continue to develop and refine the ideas, trying to ready them for an opening. The coming months and years are going to present such an opportunity.

As mentioned earlier, the nature of this revolution is that there will be widespread, fundamental breakthroughs. They are already happening in every area of science, technology, and society. Ultimately, for the system to operate with some stability, the big changes will migrate throughout the organism so that there is an internal consistency within the interface, communication, and operation of the subsystems of the larger network. Finding that equilibrium is the process that we will be experiencing during the near future. As breakthroughs happen in certain spaces, they will force other areas to operate differently. If they are unable to efficiently adapt, those threatened institutions or ideas or processes will fail, and new ideas, institutions, and processes will show up that fulfill the required function (e.g. economy, government), but in a way that is necessarily compatible with the big, forcing change that precipitated the whole thing.

This book provides a perspective on how parts of the system will reconfigure themselves in the near future. At the same time, it provides a window on the past that made this future possible. It tells the stories of the iconic, out-of-the-box thinkers and inventors who first visualized this time when their ideas would be part of a revolution that would change the way many billions of people think and live.

I have been waiting for this book for a long time—a single source that would allow me to both learn about these amazing men who were plugged into who knows where and thought like no one else before or since and, at the same time, gain a thoughtful, integrated perspective about how all of their amazing work could, in fact, rapidly change the world in which we live. It is a primer on one of the most powerful collections of forces that are converging to enable the emergence of a new world. It is very timely.

INTRODUCTION

Finley Eversole, Ph.D.

Focused, determined, enlightened public opinion is the most potent force in the world. It has no equal but has been little used.

DJWHAL KHUL

Dying civilisations are present in their final forms whilst new civilisations are emerging; cycles come and go and in the going overlap.

DJWHAL KHUL

Over the past hundred years, according to Thomas E. Bearden—a leading thinker in the free-energy and antigravity fields—there have been at least seventy successfully working free-energy technologies that could have replaced our use of fossil fuels and nuclear energy, met the world's energy needs, and given us a pollution-free environment. Yet every one of these technologies has been *suppressed* by the energy cartels, the U.S. Department of Energy, the U.S. Patent Office, the World Trade Organization, and businesses with vested economic interests in maintaining the status quo.

In addition to the alternative energy technologies discussed in this volume, there were others I would like to have included, but given the

1

sorry history of suppression of free and low-cost energy solutions that could address our global energy needs far into the future, the inventors were fearful of the suppression publicity might bring to their work. Simply and succinctly stated, *we know how to solve the world's energy problems and could probably do so within ten to twenty years.* But nothing is likely to be accomplished until the public is *educated* and *demands* the application of new energy solutions—many of which already exist but are hidden away in "black" military and governmental programs. We're told that the energy technology hidden in some of these programs is at least one hundred years in advance of anything to which the public has access! Yet it is *our money* that has paid for these revolutionary energy technologies. They are rightfully ours!

A civilization based on limited resources cannot endure. It is time we focused on technologies that will give us access to the *infinite energy* hidden in nature and available to us everywhere on the planet. Wind and solar energy technologies have their part to play, but just over the horizon are other, far less expensive technologies that do not depend on seasons and locations, do not scar the environment, and can meet *all* of humanity's energy needs for as long as the Earth exists. The public needs to know of these technologies and demand that they be implemented.

Using sound alone, the late-nineteenth-century Philadelphia inventor John Worrell Keely tapped into a limitless source of energy—apparently what we today call the zero point energy field. In the early twentieth century, Djwhal Khul made the following prediction:

> The study of sound and the effect of sound . . . will put into man's hands a tremendous instrument in the world of creation. Through the use of sound the scientist of the future will bring about his results; through sound, a new field of discovery will open up; the sound which every form in all kingdoms of nature gives forth will be studied and known and changes will be brought about and new forms developed through its medium. One hint only may I give here and that is, that the release of energy in the atom is linked to this new coming science of sound.

This brings me to the next important point: we are now moving into the Aquarian Age—a 2,100-year period during which a shared *group consciousness* and growing sense of our One Humanity will steadily replace the old "leader-and-followers" mentality of the Piscean Age we are rapidly leaving behind. The demand for democracy and freedom from dictators that gave birth to the Arab Spring and the growing demands for economic justice on the part of the Occupy Wall Street movement are examples of this emerging group consciousness. Increasingly, millions of human beings will together demand enlightened change, leading to the freedom, justice, and welfare of all our fellow citizens on Earth.

In addition to the Occupy Wall Street movement, with its demands for economic justice, we are already seeing the Occupy movement spread to other areas of concern, such as the Occupy Our Food movement, demanding farmers' rights and opposing the spread of dangerous genetically modified foods; the Occupy Education movement, demanding improvements in our educational systems; and the Occupy Our Environment movement for environmental protection.

With this volume, *Infinite Energy Technologies,* we're calling for an Occupy Our Energy Solutions movement to demand the research and application of new energy technologies that can radically transform our global economy by providing everyone on Earth with unlimited energy, drastically reduce global poverty and allow the rise of poor nations, protect our environment with pollution-free sources of energy, end political and military conflicts over limited resources, and bring about a world of peace, stability, and economic justice. All this is possible but will only come about when enough people *demand* it.

The emerging *group consciousness* we've seen in the Arab Spring and Occupy movements provides us with a first clear glimpse into the workings of a universal law that will be increasingly employed by humanity during the coming age—an age I like to think of as *the creative age* as it will be an age of creative discoveries and creative outpourings unlike anything heretofore seen on Earth. No important field of human endeavor will be unaffected.

Known to many as the law of invocation and evocation, what this law teaches us is that both Nature and Spirit respond to humanity only when we *make demands* on them. When we seek to carry out our hopes and dreams and demand the best in ourselves—and this applies equally to us as individuals and to humanity as a whole—a response will always be forthcoming, one that corresponds in nature and potency to our demands. Where no demands are made, no responses are forthcoming. Because humanity is naturally invocative, when large numbers of people are united in making *the same demands,* the potency of their united invocative appeal is greatly magnified, and so are the results. This law is an example of the working of the law of attraction. One law always holds true: *like attracts like.*

This law of invocation and evocation is predicated on an equally fundamental truth: *energy follows thought.* I like to quote Oscar Wilde, who said, "To think a thing is to cause it to begin to be." The end of the cold war in 1991 provides a good example of this process. President Ronald Reagan took credit for ending the cold war, attributing it to his massive U.S. military buildup, which the Soviet Union could not match. However, for three consecutive years prior to the end of the cold war, hundreds of thousands, perhaps millions of people all across the United States and in Europe met in groups on New Year's Eve and prayed and meditated simultaneously, calling for an end to the cold war. I took part in one gathering in Atlanta and two in Milwaukee. All *creative processes* involve three stages or the interactions of three energies. Shortly after the third gathering, in which the *mass consciousness* of hundreds of thousands of members of the human family *invoked* an end to the war, it was suddenly over. Since *like only gives birth to like,* I've always believed it was this outpouring of a demand for peace by so many people on two continents, and not Reagan's military buildup, that brought an end to the cold war.

Here you have an example of *group consciousness* and the law of invocation and evocation at work. The Arab Spring and Occupy movements are the most recent examples of the working of this law. As the law of invocation and evocation is better understood, it will be systematically employed by humanity at select times of the year to effect positive global change along predetermined lines. Social media and the Internet will allow us to synchronize our efforts. The result will be a rapid acceleration of global change in line with the hopes and aspirations of *all* peoples. The more understanding we have of what's possible, the more rapidly change will come—as with the solving of our energy problems. Our first requirement is knowledge, followed by the demand for these new energy technologies. Once these criteria are employed, we can hope for rapid, positive global change beneficial to everyone. The 99 percent *can* truly change the world! We only need to agree on what that vision of the world should be.

A bird's-eye overview of the chapters in this volume will aid prospective readers and users of *Infinite Energy Technologies* to quickly grasp its broad, innovative themes. Each author is a leading expert in his or her field. The solutions to major global problems presented here are not those with which you are likely familiar—those reported in the media and debated by politicians and big business. In these chapters we are at the frontiers of thought, discovery, and invention.

BACK TO THE FUTURE: THE LEGACY OF THE VISIONARIES

The next five chapters focus on five of the great geniuses of the world who could have taken humanity much further into the future than we have chosen to go. Some of their work has been lost or ignored, much has been classified or suppressed, and their names are all but forgotten in the annals of science. By taking another look at their legacies and restoring to them the stature they deserve, we take a small step toward reclaiming our own dignity and recovering the greater destiny that they foresaw and wanted for humanity.

Marc J. Seifer, in "Nikola Tesla," shows us the mind of the extraordinary genius to whom we owe most of the electrical inventions of the world we live in—even cell phones and the Internet. We will also read about the efforts made by his enemies to subvert or claim credit for his world-transforming achievements while doing their best to remove his name from the historical records—and very nearly succeeding.

Theo Paijmans, in "John Worrell Keely," provides a brief biography of this mysterious late-nineteenth-century Philadelphia inventor, the first pioneer of free energy and an explorer of the hidden power of music, said by some to have been hundreds or thousands of years ahead of his time. The author of the only biography of Keely, Paijmans updates his book here with new information and newly discovered photos.

Callum Coats, in "Viktor Schauberger," gives us a fascinating study of the many secrets of water uncovered by this Austrian water pioneer. These include the significance of water temperature, how flowing water behaves and why, the implications of vortex motion for reengineering our world, and the damaging effects of modern water engineering and water treatment on our environment and health.

Gerry Vassilatos, in "Royal Raymond Rife," tells the heartbreaking life story of a gentle and brilliant man who found the cure to cancer and hundreds of other diseases, only to have his life's work suppressed by the American Medical Association and the U.S. Food and Drug Administration. His work still has revolutionary implications for the future of human health without the use of drugs and surgery—hence its suppression.

Jeane Manning, in "T. Townsend Brown," introduces a physicist whose breakthroughs with flying disc airfoils in the 1950s pointed to fuel-less space travel—this brilliant man's gift to humanity. Instead, the antigravity technology known as electrogravitics was classified and hidden. We again have the story of revolutionary advances in knowledge ignored and suppressed by the scientific establishment.

INFINITE ENERGY:
A NEW SCIENCE FOR A
POLLUTION-FREE WORLD

While public debate about alternative energy has focused primarily on wind and solar power, far more dramatic and far-reaching solutions already exist or are in exploratory stages. The public knows little about these alternative energy sources, which energy cartels and governments have intentionally and sometimes brutally suppressed. These newer solutions, such as energy extracted from water, energy captured from the background energy of the cosmos (called zero point energy, the vacuum, or etheric energy), energy generated by magnetic fields, vortex energy, and cold fusion—while often defying the precepts of familiar physics—offer far more complete and long-term solutions to our energy problems. And all are locally available and environmentally friendly, producing no pollution.

Brian O'Leary, in "The Sustainable Technology Solution Revolution," sees the roots of ecocide and genocide in the control of the economy by the power elite. He urges greater public knowledge of and support for free energy and clean water technologies, which could deliver far-reaching global changes very rapidly to our world. He also appeals for an Age of Wisdom to replace our current age of greed.

Jeane Manning, in "Power for the People—from Water," surveys the leading breakthrough technologies in the free-energy field based on extracting power from water. Inventors and a few scientists are learning to release large amounts of useful energy from small volumes of water without radioactivity or other pollutants. Some of these new systems can clean polluted waters while generating electric power.

Steven Greer's "Imagine a Free-Energy Future for All of Humanity" presents us with evidence of what a real future for the Earth can look like with limitless free nonpolluting energy available at the location of use, giving humanity the ability to end global poverty once and for all. He also points out that antigravity technology, which already exists in publicly financed black programs, would make interstellar travel a reality for humanity.

Theodore Loder, in "Energy Technologies for the Twenty-first Century," reviews major research into antigravity technology, including testimony from insiders to highly secret military-industrial programs who say we *already* have advanced antigravity vehicles reverse-engineered from UFOs. Also citing zero point energy as a source of non-polluting energy that will permanently free humanity from fossil fuels, he writes, "Because antigravity and its allied zero point energy technologies potentially offer the world a future of unlimited, non-polluting energy it has been suppressed" because free energy is seen as a threat to the global economy (read, "the energy cartels").

John A. Thomas Jr. and John R. R. Searl, in "Harnessing Nature's Free Energy," detail the discovery and working of John's Searl Effect Generator (SEG), a unique magnetic device that extracts free energy from space continuously, generating electricity. Since the 1950s, the SEG has awaited public demand for its use. At high speeds, it creates its own gravitational field and becomes a gravity-free device usable for space travel.

Edmund Storms, in "Cold Fusion," tells us the reality of cold fusion has now been established beyond a doubt and represents a new phenomenon in relation to the chemical environment and nuclear reaction. Delays in research due to its initial rejection mean there are still problems to be solved, but cold fusion promises to be a valuable source of low-cost energy in the future.

Thomas Valone, in "Zero Point Energy Can Power the Future," gives us a detailed scientific look at the current state of research into zero point energy as a source of free energy. Zero point energy, also known as the energy of the vacuum state, is the background energy of the universe that, as Valone notes, "breathes life into every atom, sustaining its size and shape and it certainly is renewable."

Infinite Energy Technologies is what I hope will be the first in a series of volumes under my editorship that seek to bring before a larger public answers and solutions to many of our global problems in the areas of energy, the economy, the environment, health, food, agriculture, global change, and designs for the future.

Several well-known authors have been calling for "a new story" capable of guiding humanity through the profound global transition we have entered, preparatory to the birth of a new age and civilization, a higher consciousness, and a deeper social interconnectedness. An essential part of this story for me is the Return of Prometheus and the *liberating of the creative energies of the coming age*. As I glance into the future, I see ahead a long unfolding age of creative outpouring and accomplishment in every field of human endeavor, dwarfing in comparison all the great achievements of humanity's past. In Greek mythology, Prometheus is said to have stolen fire from heaven as a gift to humanity and, in so doing, given birth to human civilization and all the arts and sciences. For this crime against the gods, Prometheus was chained to a rock and made to suffer by having his liver devoured daily by an eagle. Eventually, however, a human hero— Hercules—succeeded against all odds in setting Prometheus free. We're never told what Prometheus did following his liberation.

The myth of Prometheus is a fitting one for our time. Today, we can retell his story as that of a liberated and wiser Prometheus coming to the aid of a suffering, yet more evolved humanity. The new Prometheus is arriving just as the human family is entering a period of profound transformation, faced with crises on all sides and the need for a new awakening of its creative and spiritual powers. It is such a time as ours that calls for Prometheus' much-needed creative fires. As we take charge of his divine fire, we are also called to the heroic and Herculean task of rebuilding civilization along new and wiser lines, inspired by a new vision of compassionate inclusiveness, which recognizes that humanity is becoming and forevermore will be One Humanity—unified in consciousness, though richly diverse in its creative contributions to the rebirth of the world.

Perhaps these new "fires of heaven" stand for the fires of freedom liberating us from false gods, outworn beliefs, and faltering institutions. They are above all fires of purification and transformation. Perhaps they also represent the discovery of the new energies needed to rebuild the world that a more awakened science will increasingly unleash on Earth in the coming age. These Promethean fires are a sacred trust, not to be exploited for selfish ends.

This inflow of new creative fire means we can re-create our world based on a wiser vision of what it means to be truly human. The liberation of the creative imagination is an essential part of our "new story." *Infinite Energy Technologies* picks up one thread of this story and gives us a glimpse into humanity's future. It offers us a new energy paradigm and a new vision of a pollution-free world, if right choices are made. Our destiny depends on choosing wisely.

PART 1

Back to the Future

THE LEGACY OF THE VISIONARIES

The resistance to a new idea is in proportion to the idea's importance.

BRIAN O'LEARY

Let the future tell the truth and evaluate each one according to his work and accomplishments. The present is theirs; the future, for which I work, is mine.

NIKOLA TESLA

I know that most men, including those at ease with problems of the greatest complexity, can seldom accept even the simplest and most obvious truth, if it be such as would oblige them to admit the falsity of conclusions which they have delighted in explaining to colleagues, which they have proudly taught to others, and which they have woven thread by thread into the fabric of their lives.

LEO TOLSTOY

Most qualified scientists shy away from things that are not generally accepted, and there is justification for their viewpoint. They are not frontiersmen. They are maintenance men for the existing order.

TREVOR JAMES CONSTABLE

We undoubtedly have in our brains some finer fibers which enable us to perceive truths which we could not attain through logical deductions, and which it would be futile to attempt to achieve through any willful effort of thinking.

NIKOLA TESLA

The intuition which guides all advanced thinkers into the newer fields of learning, is but the forerunner of that omniscience which characterizes the soul.

DJWHAL KHUL

Here's to the crazy ones. The misfits. The rebels. The trouble-makers. The round heads in the square holes. The ones who see things differently. They're not fond of rules, and they have no respect for the status-quo. You can quote them, disagree with them, glorify, or vilify them. But the only thing you can't do is ignore them. Because they change things. They push the human race forward. And while some may see them as the crazy ones, we see genius. Because the people who are crazy enough to think they can change the world, are the ones who do.

THINK DIFFERENT, APPLE COMPUTER AD

All men dream, but not equally. Those who dream by night in the dusty recesses of their minds, awake in the day to find that it was vanity; but the dreamers of the day are dangerous men, for they may act their dream with open eyes, to make it possible.

T. E. LAWRENCE

The one requirement coupled with the gift of truth is its use.

RALPH WALDO EMERSON

1

NIKOLA TESLA

ELECTRICAL SAVANT

Marc J. Seifer, Ph.D.

The scientific breakthroughs of Albert Einstein, the inventions of Nikola Tesla, the great ventures of Andrew Carnegie's U.S. Steel, and Sergey Brin's Google, Inc.—all this was possible because of immigrants.

PRESIDENT BARACK OBAMA, JULY 1, 2010

The day when we shall know exactly what "electricity" is, will chronicle an event probably greater, more important than any other recorded in the history of the human race. The time will come when the comfort, the very existence, perhaps, of man will depend upon that wonderful agent.

NIKOLA TESLA, 1893

Time makes more converts than reason.

THOMAS PAINE

On July 10, 2006, the scientific community celebrated Nikola Tesla's 150th birthday with a special conference in Belgrade, Serbia, and a new Tesla statue designed by Les Drysdale. It was unveiled at Niagara Falls

13

Figure 1.1. The Tesla statue at Niagara Falls, unveiled July 10, 2006

on the Canadian side, near Horseshoe Falls. Fifty years earlier, on Tesla's 100th birthday, this great Serbian inventor was posthumously awarded his finest honor when the International Electrotechnical Committee designated the name "tesla" as the measure of magnetic flux density (MRIs are measured in teslas). From that point on, Tesla could stand beside such other scientific giants as Ampere, Angstrom, Curie, Fermi, Hertz, Volta, and Watt.

TESLA'S EARLY LIFE IN CROATIA

Nikola Tesla was born in 1856, in the heat of the summer during a lightning storm, about twenty miles east of the Adriatic Sea in the small hamlet of Smiljan in Croatia. Located along a great plane with the Dinaric Alps in the distance, the Tesla home and nearby church sat against a large protective hill. This site enabled young Niko to view many lightning storms as they rode along the Alps or swept across the plain to engulf the hamlet.

The son of a Serbian Orthodox priest, Tesla was the fourth of five children, three girls and two boys, with his brother Dane (pronounced "Dah-nay"), seven years older, being the oldest child. Tesla's father was descended from military and clergymen, and his father's brother was a mathematics professor; his mother, whose family name was Mandic, came from a more prominent family. Her maternal grandfather was given the French Medal of Honor by Napoleon himself in 1811. One of her brothers was a field marshal in the imperial Austro-Hungarian Army, and she had an Uncle Petar Mandic, who was the regional bishop of Bosnia.

Figure 1.2. The home and farm in Smiljan where Tesla was born

Great things were expected of the older brother, Dane, who, like young Niko, was gifted with a photographic memory and great powers of visualization. A creek ran by their house and small farm, and Dane taught Niko how to build a water wheel in it. He also taught him how to shoot crows with a popgun. These two children's toys would become the seeds of two of Tesla's inventions: (a) the great hydroelectric turbines and power system that, even today, provide a vast percentage of the clean, renewable electrical energy for the world, and (b) Tesla's controversial particle beam gun, a secret weapon given to several countries, including the United States, just as World War II was beginning.

In about 1861, a great tragedy befell the Tesla family when Dane died in an accident with the family horse, an Arabian steed that had saved the father one previous winter in a run-in Mr. Tesla had with a pack of wolves. This event caused great upheaval within the family, so much so that they moved several miles away to the bustling town of Gospic, where Niko, now about five, could begin his schooling.

This event was important for many reasons. It caused the family to move away from their idyllic farm and probably created a rift in Tesla's relationship with his mother, who no doubt favored Dane for the simple reason that he was the first child, her first son, and a youngster she knew for seven long years before Niko was even born. Later in life, in 1899 when Tesla expressed great interest in contacting extraterrestrials, it is possible that subconsciously he was also trying to connect with his long-dead brother.

Another important aspect about Dane's death is the role of rumor and disinformation in any discussion of Tesla's life and impact on the world. Certain articles and biographies suggest that the five-year-old Niko pushed Dane down a staircase and thus killed him. It took me a while to track down the source of this rumor. It came from Arthur Beckhard, a writer whose screenplays included *West Point of the Air, Our Little Girl,* and *Sky Parade.* In 1959, Beckhard wrote the children's biography *Nikola Tesla: Electrical Genius* (published by Messner Books), which included many vignettes that he simply made up. However, this book filtered its way to the Tesla Museum in Yugoslavia, where the tale was picked up by Serbian writers, and that fictional source was used by other biographers and critical writers of Tesla who thought they had an inside scoop.

TESLA'S BREAKTHROUGH
WITH AC CURRENT

In 1878, Tesla attended the University of Graz in Austria, and two years later, the Charles-Ferdinand branch of the University of Prague in Bohemia, where, in both schools, he took such courses as geometry, integral calculus, analytical chemistry, machine construction, botany, wave theory, optics, philosophy, French, and English. At Graz, in a course on experimental physics, his teacher Professor Poeschl presented a conundrum in electrical engineering that determined his future. Poeschl showed the class an electrical turbine that used a commutator to change the naturally flowing alternating current at the source of power into direct current (DC), so that it could be transported down a wire to run a machine or light a lightbulb.

Electrical lighting for factories and homes was in its infancy, but the direct current power distribution system, which had been in use about twenty years, was highly inefficient. The problem had to do with the mystery of electricity, which by its nature was alternating, that is, it changed its direction of flow many times per second. Consider a river that is flowing downstream and then upstream and then downstream many times per second. How does one make use of such unruly power, that is, make the waterwheel go in one direction?

The trick for making electricity go into a single direction had first been accomplished by William Sturgeon in 1832 with his invention of the commutator. This was a series of wire brushes with a small gap in the middle of them. When the electricity was flowing downstream from the generator, it had the impetus to jump the gap to make it to the motor or lightbulb on the other side, and when it reversed its direction, nothing happened. In this way, the commutator captured a unidirectional flow, but it was highly inefficient.

Tesla saw in an intuitive flash that the commutator could be removed and the alternating current harnessed in an unencumbered fashion. Tesla was an A+ student, and Poeschl initially complimented the young man, but then essentially embarrassed him. "Tesla, you may do many things," Poeschl said. "But this is a perpetual motion scheme, an impossible task."[1] Dropping out of school before graduation, Tesla would spend the next four years in a daily quest to solve this puzzle and prove Poeschl wrong.

His first job was in Budapest as an electrical engineer for the Pukas brothers, associates of Thomas Edison. It was at this time, in 1882, that Tesla solved the problem of removing the commutator with his first major invention, the rotating magnetic field. Shortly thereafter, Tesla worked at the Edison electrical lighting plant in Paris and then, in 1884, moved to New York City to apprentice with the Wizard of Menlo Park himself. Unfortunately, this union would only last a year, in part because Edison was unwilling to try to understand Tesla's invention with alternating current. Tesla went out on his own, found new backers, and soon sold his entire forty-patent AC polyphase electrical power distribution system to Westinghouse for five hundred thousand dollars, plus royalties. Approximately half of this went to his partners.

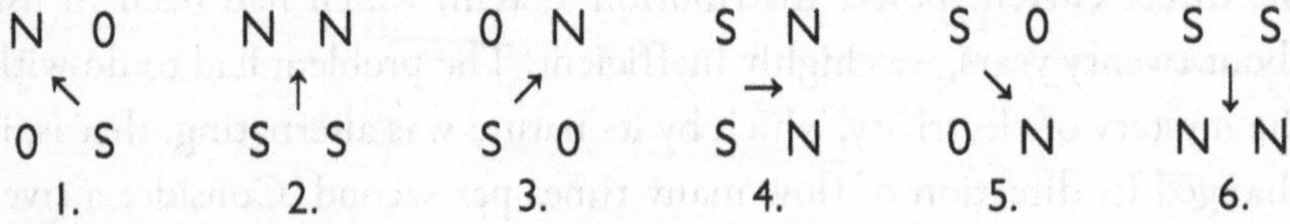

N O	N N	O N	S N	S O	S S
↖	↑	↗	→	↘	↓
O S	S S	S O	S N	O N	N N
1.	2.	3.	4.	5.	6.

Figure 1.3. Follow the north poles (N). Each quadrant represents two circuits catty-corner to each other. The north-south (N/S) pole in circuit 1 (top left corner to bottom right corner) starts to reverse itself by appearing in section 2 as still N/S, in section 3 as 0/0 (because it is reversing), and then as S/N in sections 4, 5, and 6 as it begins to reverse itself. The other circuit (top right to bottom left) is out of phase so that an armature attracted to the north pole (the arrow) will spin, thus producing the rotating magnetic field.

TESLA'S CONTRIBUTIONS TO HYDROELECTRIC POWER

Edison, and his competitors Elihu Thomson of the Thomson-Houston Company and George Westinghouse, were dotting the landscape with coal-operated electrical power stations at every other mile across cities and towns, because a mile was as far as electricity could travel at that time. DC generators and competing AC generators, which also used commutators, could only be used for lighting homes, *not* for running electrical equipment, unless these generators were placed inside or right next to the actual factory.

Before Tesla's invention of the rotating magnetic field, all factories were situated along rivers, but after Westinghouse purchased his patents in 1889, factories would soon be able to be located anywhere as long as a power line could reach them. By setting up two or more electrical currents out of phase with each other, Tesla made the commutator obsolete and eliminated the need for the thousands of polluting coal-operated power stations. Tesla's clean energy invention was renewable because it ran on waterfalls. The full creation took forty patents to describe and included the now ubiquitous induction motor. This fundamental creation, unchanged today, over one hundred years later has saved literally billions and billions of man-hours because machines were able to replace the work that man had done in the past. Tesla, in fact, realized this, and therefore predicted that with the advent of his induction motor and AC electrical power distribution system, society would advance at a more rapid rate because humans could spend more time in study rather than in manual labor.

With the present discussion of potentially resurrecting nuclear power, let's keep in mind the comparisons between it and hydroelectric power. A clean energy, nonpolluting, renewable Tesla hydroelectric plant situated at Niagara Falls provides electricity to tens of millions of outlets, including homes, stores, factories, and streetlights hundreds of miles away. The power from Niagara Falls powers and illuminates parts of Canada, large and small cities throughout Maine, Vermont, New Hampshire, Rhode Island, Connecticut, New York, and even Pennsylvania, and also such great metropolises as Toronto, Montreal, Boston, Providence, Hartford, and New Haven. It also powers Manhattan and all of its major boroughs such as Brooklyn, the Bronx, and Queens.

Compare this to a nuclear power plant, which poses a tremendous safety threat: There is no place to store the spent radioactive material. There is also the potential that an accident or terrorist act could wipe out thousands or millions of people. Additionally, such a plant can only supply, at best, about a tenth of what a hydroelectric plant can. Finally, one has the problem of radioactive leakage, with its attendant risk of genetic mutations and cancer. The only goal of a nuclear power plant is simply to boil water to create steam—in other words, to simulate what a waterfall or hydroelectric system does naturally!

TESLA'S IMPACT ON THE WORLD

Tesla spent his life seeking alternative forms of renewable energy sources and wrote several treatises on solar, tide, wind, and geothermal power. One of his more creative ideas was based on W. H. Wellaston's (1766–1828) invention of a dual-exhausted bulb called a cryophoros, which allows liquids to boil at room temperature. Tesla's plan, published in the article "Sea Power Plant Designed by Tesla" in the *New York Times* on November 11, 1931, was to construct a gigantic Wellaston dual-bulb system in the oceans. Using the differences in temperature between the surface and the great depths, a cycle of heat transfer would be created inside large vacuum bulbs that would boil the oceans at low temperatures so that the steam could be used to drive turbines to create electricity.

In 1892 Tesla spoke before the Royal Society of London. Present were such scientists as Sir William Crookes (inventor of the Crookes tube), J. J. Thomson (discoverer of the electron, winner of the Nobel Prize in 1906), Nobel nominee Oliver Heaviside, Sir John Ambrose Fleming, Sir James Dewar (inventor of the thermos bottle), Sir Oliver Lodge (inventor of wireless technology), Lord Rayleigh (winner of the Nobel Prize in 1904 for discovering argon), and Lord Kelvin, who said, "Tesla has contributed more to electrical science than any man up to his time."[2]

Other Nobel laureates who were either influenced by Tesla or congratulated him in 1931 for his seventy-fifth birthday included Albert Einstein, Robert Millikan (who had attended Tesla's 1891 Columbia University lecture), Arthur Compton (who received the prize for his work in optics, X-rays, and high frequency phenomena), and Ernest Rutherford (whose planetary model for the atom—with Niels Bohr—is based on information from Tesla's Royal Society talk and its related Columbia University lecture).[3]

This treatise of one hundred pages, which combined the 1891 Columbia University and 1892 Royal Society of London lectures, was published in the electrical journals of the day, and in Tesla's collected works published two years later by *Electrical Engineer* editor T. C. Martin.[4] During these lectures, Tesla likened "the infinitesimal world, with molecules and their atoms spinning and moving in orbits" to "celestial bodies spinning with them ether, or in other words, carrying with them static charges."[5]

This model, through the work of Ernest Rutherford and Niels Bohr, would go on to become the standard for the atom and win for this duo a Nobel Prize, but Tesla would receive no credit. (One of Ernest Rutherford's favorite books, as noted in Orrin Dunlap's *Radio's 100 Men of Science*,[6] was *The Researches, Writings, and Inventions of Nikola Tesla*.) According to NobelPrize.org, Rutherford did much of his early work on "magnetic properties of iron exposed to high-frequency oscillations," precisely the kind of work inaugurated by Tesla in his early lectures in the 1890s. Further, Niels Bohr spoke in praise of Tesla at the 1956 centennial congress to celebrate Tesla's birth.[7]

Tesla has also been written out of books on the history of physics even though his treatise became a standard, and his work became essential for the study of matter. (Super colliders rely heavily on the Tesla coil and other Tesla technology.)

As a finale to Tesla's 1891–1893 electrical lectures, and as a way to tell the world that alternating currents were perfectly safe if used correctly, Tesla sent hundreds of thousands of volts through his own body to issue forth sparks from his fingertips and illuminate wireless cold lamps. Part of Tesla's reason for sending electricity through his

Figure 1.4. Tesla sending five hundred thousand volts through his body in 1898 to illuminate a fluorescent tube. This photo was taken with a strobe as the inventor whipped the tube around.

body was because Edison, still stuck in direct current, was engaged in a publicity campaign to suggest that AC was more dangerous. Edison even went so far as to talk New York State into instituting an electric chair using AC current to put heinous criminals to death, or in Edison's words, "to Westinghouse them." As is well known, as part of this negative publicity campaign, Edison began electrocuting stray dogs and cats and even a rogue elephant with the horrible Tesla/Westinghouse currents.

TESLA'S DEAL WITH WESTINGHOUSE

Westinghouse was in a bind with his backers. At the time he purchased the Tesla AC polyphase system, Westinghouse had nearly one thousand profitable AC electrical plants throughout urban America. Like his competitors, Edison Electric, which used DC, and Thomson-Houston, which like Westinghouse used AC, Westinghouse still made use of a commutator. Thus, their plants could not power home appliances. They were used exclusively to illuminate homes or apartment complexes within a radius of approximately twenty to twenty-five blocks, or on site, to run equipment for a single factory. To change over to the Tesla system would mean scrapping all of these profitable stations. With his eye on the larger picture, Tesla agreed to destroy the royalty clause of his contract if Westinghouse committed himself to making the change. And that is what happened. Soon after, the Chicago World's Fair of 1893 became the first metropolis to be lit by the Tesla-Westinghouse AC Polyphase System.

Once it became apparent how superior the Tesla system was, Westinghouse's competitors joined forces under the guidance of J. Pierpont Morgan (the most powerful financier of the day) to create General Electric. It was now the task of Elihu Thomson and Charles Steinmetz to figure out a way to usurp the Tesla invention but do it in a manner that somehow did not look like piracy. (That's why Steinmetz would write textbooks on the invention but do it in a way that left out Tesla's name.) Legal suits soon began and then continued to sap funds from both camps. As the years went on, electrical engineering students,

Figure 1.5. Nikola Tesla, circa 1895

relying on the textbooks of Steinmetz, would come to believe that he and not Tesla was the author of the AC polyphase system.

This battle of the currents ended abruptly when Edison left his company to begin a career in filmmaking. This occurred right before Tesla spoke at Niagara Falls in 1897. Since the harnessing of the falls was such a huge endeavor, Westinghouse simply had to share the work with GE. Westinghouse obtained certain trolley patents in return, and that's how General Electric obtained access to Tesla's patents without paying the inventor a cent. At the same time, the combination of these new train patents and the Tesla AC polyphase system allowed Westinghouse to launch an entirely new endeavor as well: the electric subway. Again working together with Morgan, this subway system was installed in New York City right after the turn of the century, and again, Tesla received absolutely no recompense.

TESLA ACHIEVES GREATER RENOWN

Tesla was quite wealthy from the original deal with Westinghouse and thus in 1897 was able to move into the Waldorf Astoria and to move his laboratory to Houston Street. It was at this time, during the height of the Gay Nineties, that Tesla befriended Robert Underwood Johnson, the editor of *The Century* magazine, and his alluring wife, Katharine. Introductions had been made by T. C. Martin, editor of both the prestigious *Electrical World* and the groundbreaking book *The Researches, Writings, and Inventions of Nikola Tesla*.

Through the Johnsons, Tesla met many of the greatest writers, artists, and dignitaries of the day, most of whom were invited back to the wizard's laboratory, where he could display his latest creations. This group included John Jacob Astor and his wife, Ava Willing, architect Stanford White, writers Mark Twain and Rudyard Kipling, the conservationist John Muir, Teddy Roosevelt and his sister Mrs. Douglas Robinson (one of the founders of the Metropolitan Museum of Art), Mary Mapes Dodge (author of *Hans Brinker*), the playwright Marguerite Merrington, composer and conductor Anton Dvorak (who wrote the *New World Symphony* at this time), and Ignace Paderewski (the most successful pianist of the day, and latter-day prime minister of Poland).

In 1897, the Tesla hydroelectric system was installed at Niagara Falls, where Tesla spoke as the inventor. It is virtually unchanged today. During his Niagara Falls speech, Tesla discussed his next invention, the transmission of electrical power without wires. The seeds of this creation were displayed in highly publicized lectures that Tesla gave throughout the early 1890s at the Royal Academies of Science in London and Paris and at electrical congresses in St. Louis, Chicago, Philadelphia, and New York. So astonishing were these lectures that thousands of people came to hear them.

In 1891, speaking at Columbia University before such peers as Elihu Thomson (later head of General Electric), Alexander Graham Bell, Michael Pupin (professor of physics at Columbia), Robert Millikan (latter-day Nobel Prize winner on his work with cosmic rays), and gyroscope inventor Elmer Sperry, Tesla displayed his work in high-frequency phenomena. Tesla began his talk somewhat nervously, but gained momentum as it progressed: "Of all the forms of nature's immeasurable, all-pervading energy, which ever and ever change and move, like a soul animates an innate universe, electricity and magnetism are perhaps the most fascinating. . . . We know," Tesla continued, "that [electricity] acts like an incompressible fluid; that there must be a constant quantity of it in nature; that it can neither be produced or destroyed . . . and that electric and ether phenomena are identical."[8] Having set the premise that our world is immersed in

Figure 1.6. Mark Twain in Tesla's laboratory, circa 1894

a great sea of electricity, the wizard proceeded to astound the audience with his myriad experiments.

At this time, Tesla also displayed his Tesla coil, which stepped up frequencies to very-high voltages, precursors to both radio and television tubes, a button lamp that could disintegrate matter, and fluorescent and neon lights that illuminated when the proper resonant frequency was generated. Before this moment, the world was caught up in Heinreich Hertz's wireless spark-gap device. This Hertzian apparatus was later advanced by Guglielmo Marconi to transmit Morse code over long distances. But the problems with this device were threefold: it was next to impossible to establish separate channels, weather conditions could easily interfere with transmissions, and complex forms of information could not be transmitted.

TESLA GOES WIRELESS

On the other hand, Tesla's invention of *continuous wave* wireless communication, in combination with high-frequency phenomena, and aerial and tuned circuits, lay the groundwork for our modern age of cell phone technology, remote control, wireless transmission, and mass communication. In particular, Tesla's use of the ground connection paved the way for radio as well as wireless transmission of text and pictures to newspapers and also TV transmission. His vacuum tube, which responded to wireless messages sent from distant places, was constructed on principles he learned from studying the human eye, as both eye and radio tube responded to signals sent from afar.

In 1904, Arthur Korn, an electrical engineer from the University of Munich, gained the attention of the scientific community when he successfully transmitted photographs sent along wire transmission lines from Munich to Nuremburg. The ability to send such complex information by means of electricity was, at the time, an astonishing accomplishment, and it was eclipsed just a few years later when Korn accomplished this without wires! By 1913, Korn was also able to send wireless photographs across the Atlantic. According to Korn, who is often credited as the first inventor of the television tube, the apparatus used "Tesla currents." In 1920, Tesla explained in an *Electrical Review* article titled "The Art and Practice of Telephotography" that Korn's tube, which was based on a tube Tesla displayed in the early 1890s, "is excited by a high-frequency current supplied from a Tesla transformer and may be flashed up many thousand times per second,"[9] thereby attaining the moving television image. That's precisely how the TV tube works. (Philo Farnsworth, often credited as the inventor of the TV tube, was fourteen years old in 1920.)

Tesla's work in wireless technology took a veritable quantum leap forward in 1898, when he displayed his remote-controlled robotic boat, which he called the "telautomaton," in the electrical journals and before the public at Madison Square Garden.

Tesla realized that if, for instance, a remote-controlled torpedo were to be launched, it would be a simple matter for the target ship to send out its own signal to cause the weapon to turn around and attack the hand that sent it. Having studied Herbert Spencer's work on nerve conduction, Tesla got the idea of *combining frequencies* to send complex information along separate secure channels. For instance, say one only had ten wireless frequencies to work with (if we are talking about cell phones, this means that a manufacturer could only create ten separate channels and thus be able to sell only ten separate cell phones). However, if combinations of frequencies were used, the amount of potential channels would increase in geometric proportions: combining two frequencies would create ten times ten, or one hundred channels, three frequencies, a thousand channels, and so on.

This invention, which Tesla patented in 1901, became the basis for protected privacy communication and radio-guidance systems developed by his protégé John Hays Hammond Jr., who called the invention Tesla's "prophetic genius patent." By combining frequencies, Tesla had set the stage for the age of cell phones, whereby a virtually unlimited number of individual wireless phone numbers could be set up.

But the telautomaton was much more than this. Since the inventor could send a signal to the boat and the boat would respond, from Tesla's point of view he had created the first prototype of a *thinking* machine, the first of a new species on the planet, "not made out of flesh and bones, but rather of wire and steel." From the construction of Tesla's telautomaton came such devices as the garage door opener, the wireless car-lock system, cell phone technology, encryption devices, the TV remote, wireless communication, radar, artificial intelligence, and robotics.

Figure 1.7. Tesla's remote-controlled robotic boat, displayed at Madison Square Garden in 1898

Having astonished the world in four separate areas: electrical power transmission, fluorescent lighting, wireless communication, and robotics, and funded by John Jacob Astor, Tesla set out in 1899 to Colorado Springs to experiment with the transmission of electrical power to distant points without the use of wires. His idea was simple: use the resonant frequency of the Earth itself as a carrier wave to distribute electrical power. Having constructed a prototype planet, Tesla calculated the length of his electrical frequencies and where the nodal points would be. At these junctions around the globe, the inventor hoped to eventually erect receiving towers. In that way, if, for instance, a major wireless transmitter were put in at Niagara Falls, electrical power could be transported through the Earth and the air to such distant places as the Sahara Desert or the upcoming Paris Exposition that was planned for 1902.

Figure 1.8. Tesla's wireless transmission tower in Colorado Springs

Tesla erected a two-hundred-foot-tall transmission tower in Colorado, and from this experimental station he created sixty-foot-long lightning bolts and measured the resonant frequency of the Earth. Tesla also claimed that he circumscribed the Earth with his electrical waves, setting up a nodal point at the antipode, which he calculated was in Australia. After tracking thunderstorms at distances of six hundred miles Tesla announced to the world in a *Colliers* article titled "Talking with the Planets" that he had received pulsed frequencies from outer space. These, he speculated, came from intelligent beings from some neighboring planet like Venus or Mars.

A year later, Tesla returned to New York City and his home at the Waldorf Astoria. At this time, he formed a partnership with J. Pierpont Morgan so that he could finance the creation of a wireless transmitter to send signals across the Atlantic. Having leased a 1,800-acre tract out on Long Island from a Mr. Warden, Tesla called his planned radio city Wardenclyffe. His architect was Stanford White, who was the designer of the capitol building in Providence, Rhode Island; Rosecliffe Mansion and the Tennis Hall of Fame in Newport; the Washington Arch in Greenwich Village; Madison Square Garden in New York City (which was funded by Morgan); the Agricultural Building at the Chicago World's Fair; the Player's Club in New York; and numerous other buildings.

Figure 1.9. Political cartoon of J. Pierpont Morgan

Morgan had given the inventor $150,000 to complete the ninety-foot tower and laboratory. But just as Morgan sailed for Europe for his annual extended stay, Tesla learned that Marconi was pirating his apparatus. Thus, unbeknownst to his benefactor, Tesla decided to double the size of the tower. His thinking was that by doing this, not only could he send messages to Europe, but now he could reach the whole world as well. Thus, when completed, instead of the revenues being merely doubled (because the costs had doubled), they would increase in geometric proportions because everybody on the planet would be tuned into his system.[10]

Figure 1.10. Guglielmo Marconi (fourth from left)
with his wireless device

Marconi's success in transmitting a Morse-coded signal across the Atlantic in 1901 captured the imagination of the world and hurt Tesla's credibility with Morgan. Where Tesla was envisioning the transmission of voice, pictures, and power, and the creation of an unlimited number of wireless channels (equivalent to today's cell phone technology and achieved by using multiple frequencies), Marconi could only send pulsed signals on a very small number of channels and admittedly had no under-standing of selective tuning. There was no comparison between the two inventions whatsoever from a technical point of view. However, Tesla's ideas were so futuristic that Morgan simply couldn't comprehend them. Stop and think for a moment yourself how our cell phones transmit voice and pictures through the air. Unless you really study the mechanism involved, it does indeed seem like magic. In a series of articles and in numerous letters, Tesla pleaded with Morgan to give him the funds to complete the tower and laboratory.

Figure 1.11. Franklin Roosevelt

Tesla would later have another competitive brush with Marconi. When World War I was in full swing, Guglielmo Marconi chose that time to sue the U.S. Navy for patent infringement for using his wireless apparatus. Assistant secretary of the U.S. Navy, Franklin Roosevelt, researched the problem and found that at the turn of the century Nikola Tesla had been negotiating with the U.S. Navy, via the Light House Board, to place his own wireless apparatus on naval vessels. This correspondence also explained that Tesla's work predated and was superior to the work of Marconi. Thus, it established Tesla's priority.

Consequently in 1916, during the Marconi/U.S. Navy litigation, Roosevelt wrote that, based on Tesla's 1899 correspondence with the Light House Board, the U.S. government now had "suitable proof of priority of certain wireless usages by other than Marconi."[11] This was later used by the U.S. Supreme Court to rule in Tesla's favor as the original inventor of the radio.[12]

THE FAR-REACHING EXTENT OF TESLA'S VISION

The following quote is from a piece that Tesla published in 1904 in the journal *Electrical World and Engineer*. It is a splendid article that has pictures of both his magnificent wireless towers, which were erected in Colorado Springs and Wardenclyffe on Long Island. The article was written specifically to explain to Morgan the inventor's vast plan. Note how Tesla, in a sense, foresees what became the Internet.

> [My] World Telegraphy System . . . involve[s] the employment of a number of plants each of [which] will be preferably located near some important center of civilization, and the news it receives through any channel will be flashed to all points of the globe. A cheap and simple [pocket-sized] device may then be set up on land or sea, and it will record the world's news or such special message as may be intended for it. Thus the entire earth will be converted into a huge brain, as it were, capable of response in every one of its parts. Such a single plant . . . can operate hundreds of millions of instruments, the system will have a virtually infinite working capacity.

Even today, Tesla's understanding of the potential for world advancement and even world peace through mass communication remains profound.

A stock market crash and a series of arguments with Morgan that culminated in 1906 destroyed all chances for the tower to be completed, and it was eventually torn down in 1917. The salvage was used as partial payment to the Waldorf Astoria for back rent owed.

In 1908, there was a tremendous explosion in Tunguska, Siberia. An area the size of Rhode Island was obliterated. No one knows for sure what happened. One creative theory initiated by Tesla researcher

*Figure 1.12. Tesla's world wireless system on
Long Island, called Wardenclyffe*

Oliver Nicholsen suggests it was Tesla's doing. Tesla did indeed state in 1915, in an article in the *New York Times,* that his Wardenclyffe tower could be used as a massive Star Wars–like weapon. Under the unlikely possibility that Tesla was responsible, he would have had to have jury-rigged the tower, because Tesla's generator at Wardenclyffe was removed in 1904 by the Westinghouse Company for lack of payment.

The most likely hypothesis is that a comet or asteroid skimmed the Earth in that area and flattened out hamlets and towns and tens of thousands of trees. There is no crater and no residue from an extraterrestrial object, so a direct hit by a meteor can be discounted.

Two years later, in 1910, already concerned with preserving the world's reserves of petroleum, Tesla began work on his bladeless steam turbines, which he was hoping would replace the gasoline engine in Henry Ford's cars. This plan, which eventually was partially funded by Morgan's son, J. P. Morgan Jr., fell through, and Tesla also had a falling out with John Hays Hammond Jr., who was constructing radio-guided torpedoes for the United States, as well as for German and Japanese navies. Tesla did, however, in 1915, help the Germans construct their transatlantic wireless plants in Tuckertown, New Jersey, and Sayville, Long Island. (Once America entered the Great War, Tesla, as a patriot, severed all ties with the Germans.)

Figure 1.13. Tesla's tilt-rotor flivver plane paved the way for today's Osprey, a $40-million military airplane that takes off like a helicopter and then flies like a normal plane after the propeller is rotated 90° into the normal airplane position. Tesla's patent application drawing (top left) was published in the newspapers in the late 1920s.

Tesla then began work on a variety of flying machines. These included a hovercraft and a reactive jet dirigible, which was an early prototype that led to such airplanes as the vectored-thrust Harrier jet and the flying wing or stealth bomber. Tesla also patented a tilt-rotor aircraft that he called the flivver plane. This machine, which took off like a helicopter and then, after rotating the propeller, flew like an airplane, evolved a half-century later into the V-22 Osprey aircraft, a $40-million helicopter/airplane that the military uses today. At the time (1920s and early '30s), Tesla envisioned the vehicle competing with the automobile. Another feature of one of his airplanes was that it had no fuel on board, but rather ran on an electric motor that derived its power from ground transmitters.

TESLA'S LATER INVENTIONS

During the Roaring Twenties right into the time of the Great Depression, Tesla began work on two top-secret devices. The first was an electric car, a Pierce Arrow, which derived its power from a distant source. In 1930, Tesla met with Heinreich Jebens, a former naval officer and current director of *Deutsches Erfinderhaus* (the German Inventor's House, an organization of 10,000 inventors and researchers in Germany). Jebens was in the states to meet with Tom Edison. But a man on the ship coming over suggested he also contact Nikola Tesla. Arrangements were made and Tesla met Jebens at the Waldorf Astoria Hotel and then took the German inventor up to Buffalo to see Tesla's top-secret car under the condition that Jebens "keep absolutely quite about this" and Jebens agreed to do so.

From an e-mail that Jebens's son sent to the author (on November 15, 2003) we learn that the car was "installed with an 80 HP-electric motor," which replaced the "fuel engine" and possessed a six-foot-tall aerial and also a ground connection, which, perhaps, scraped the Earth as the car moved along. Tesla attached a converter box comprising a dozen radio tubes, "24 resistors and diversified cables," which, it must be assumed, derived power from a secret transmitter he had to have set up at the power station at Niagara Falls. After "inserting two rods into the engine," the car was ready to roll, achieving, according to Jebens, speeds in excess of ninety miles per hour. After the jaunt, Tesla directed their driver to return to the power station and "disconnect the converter, which Tesla took back to New York."[13]

The other secret invention involved the harnessing of cosmic rays, which were somehow converted into electrical power. In his twilight years, Tesla was particularly cagey about the details of this invention. In a July 11, 1937, article in the *New York Times* titled "Dr. Tesla Predicts Linking Planets," the inventor suggested this was "a different kind of energy than is commonly employed. . . . It travels through a channel of one-millionth of a centimeter." The energy, which stemmed from cosmic rays coming from the sun and other stars, traveled at many times the speed of light. According to Tesla's understanding, radioactivity was not due to the disintegration of the neutrons in the nucleus of such elements as radium or uranium, but rather, a release of cosmic energy "from all parts of the universe."

From this vantage point, radioactivity would be the end result of the inability of atoms of radioactive material to maintain their integrity. So, in a sense, they would be acting like a sieve to transfer cosmic energy into what we call radioactivity. Tesla thereby told the *New York Times* on February 6, 1932, that "if this cosmic radiation could be wholly intercepted, the radioactivity would cease."[14] If, indeed, matter is constantly absorbing ether to maintain the integrity or spin of the elementary particles, then radioactivity can now be looked at in an entirely different light. This theory also explains what Tesla called his *dynamic theory of gravity* in that the Earth is constantly absorbing energy or ether, and this is why light bends around large planetary and stellar bodies. Light is drawn in by the influx.

This theory also addresses Einstein's forty-year quest for Grand Unification, namely a way to combine gravity with electromagnetism, and explains, in its final form, what physicists of today call the God particle, the particle that gives matter its mass. According to Tesla's view, as I explain in detail in *Transcending the Speed of Light,* it is not a "particle," but rather an on-going *process* of ether absorption that explains the concept behind the God particle.*

According to Tesla, by using a special tube, he would be able to capture and transmit this unusual form of cosmic power to great distances, including to nearby planets.

Tesla's last great accomplishment, which he described as far back as 1915 but which he publicized in the 1930s, was often called the death

---

*See chapters 4 and 5 in my book *Transcending the Speed of Light* (Inner Traditions, 2008), for a comprehensive explanation via Tesla's theory of what Einstein called Grand Unification, which is directly linked to what today is called the God particle, or Higgs boson, the supposed particle that gives matter its mass.

Figure 1.14. Tesla's idea of supplying wireless power to airplanes so that they would run on electricity instead of petroleum is depicted in this futuristic drawing (slightly modified) by science fiction artist Philip Paul from about 1920.

ray. In actuality, it was a particle beam weapon that was based on principles found in the popgun that he had used to shoot crows with when he was a child.

The idea was simple. Set up in an analogous way as a Van de Graaff generator, Tesla generated a belt of highly charged ions that circulated around the base and tower of the machine. At the location where the barrel of the gun began, the inventor figured out a way to chip off small pieces of tungsten, which carried the same charge as the electronic belt. In this way, the individualized pellets would be *repelled* out of the barrel at enormous velocities, much the same way a cork would pop out of a pop gun. As World War II was brewing, it was Tesla's plan to provide these Star Wars–type weapons to all the allies, so that their borders would become impregnable from Nazi invasion. Thus, there would never be war again. Such was the mind of this great dreamer.

Tesla's life has always sparked the imagination of the novelist. He is often depicted in fictional form as a mad scientist, for instance, and has been cited as a part of the composite hero John Galt in Ayn Rand's

*Figure 1.15. Tesla in the 1930s explaining
the machinations of his particle beam weapon
to a reporter.*

Atlas Shrugged. He appeared in stories in the science fiction magazines from the 1920s and '30s produced by his friend Hugo Gernsback, and also appears in James Redfield's book *The Tenth Insight,* the sequel to *The Celestine Prophecy,* one of the most successful books ever published.

In 2006 and 2007, Tesla was mentioned on such successful TV shows as *Studio 60* on NBC and on Fox, where the words "Tesla got robbed!" appeared on the blackboard behind actor Hugh Laurie, who plays the doctor on the very popular show *House.* Photos of Tesla can be seen in such movies as *Tucker,* directed by Francis Ford Coppola starring Jeff Bridges, and *Antitrust* with Tim Robbins. Tesla also appeared on-screen in 2007 in the movie *The Prestige,* about dueling magicians from the turn of the twentieth century. Tesla was also assumed to be the real life personality behind the character of the electrical savant in the 1976 movie *The Man Who Fell to Earth* (which featured rock star David Bowie as the titular extraterrestrial and was based on the book by Walter Tevis).

Tesla's greatest acclaim of recent days can be found in the automotive industry, with the new car called the Tesla Roadster. This electric car pays homage to Tesla in that it is a real attempt to design a new clean energy technology. The Tesla Roadster is supported by former actor and California governor Arnold Schwartzenegger, who owns one, actor George Clooney, who also drives one, and such young Silicon Valley billionaires as Jeffrey Skoll of eBay and the film company Participant Media (producers of *Syriana, An Inconvenient Truth, Charlie Wilson's War*), Elon Musk of PayPal, SpaceX (the first privately owned space ship) and founder of Tesla Motors, and Sergi Brin and Larry Page of Google, who are all partners in the new Tesla automobile. Clearly, Silicon Valley recognizes Tesla as the first inventor of what is now called green technology: the design of high-tech creations that are nonpolluting and environmentally friendly.

TESLA'S GENIUS RESENTED BY PEERS

Scientists whose work is based on Tesla's inventions include: Elihu Thomson, one of the key founders of General Electric (GE), a company built on Tesla's patents for electrical power distribution; Charles Steinmetz, known for his textbooks on the AC Polyphase System, which brazenly leave out Tesla's name; George Westinghouse, whose company, like GE, was built on Tesla's alternating current (AC) invention (through the years, many Westinghouse biographers neatly removed or obscured Tesla's role in the success of that great enterprise); Elmer Sperry, whose gyroscope is based entirely on Tesla's invention of the rotating magnetic field; Guglielmo Marconi, who received a Nobel Prize for sending the first wireless message across the Atlantic Ocean in 1901, and who admittedly used a Tesla coil to achieve this end, although he claimed his work was not based on Tesla's; and Michael Pupin, a fellow Serb who taught physics for many years at Columbia University and who used other aspects of Tesla's inventions to patent a traditional method of long-distance telephonic communication over wires.

Pupin, who purposely kept Tesla's name out of electrical engineering courses at Columbia University for decades, was given a Pulitzer Prize for writing *From Immigrant to Inventor,* a book that, although discussing such things as the harnessing of Niagara Falls, goes out of its way to completely obliterate Tesla's role in the development of our modern electrical age. Instead, it places most of the credit on Pupin's friend Elihu Thomson, who, along with Steinmetz, pirated Tesla's apparatus before they legitimately obtained a license to use the AC polyphase system from the owner of the patents, George Westinghouse.

Pupin's animosity for Tesla and loyalty to Thomson and Marconi is well known. Pupin even went so far as to testify on Marconi's behalf at the aforementioned trial between Marconi and the United States Navy in 1915 whereby Marconi claimed that the navy was pirating his wireless apparatus.

One of the key sources for uncovering this blatant attempt to erase Tesla from the history books was Edwin Armstrong, inventor of AM and FM radio, a pallbearer at Tesla's funeral in 1942 and a student of Pupin in or about 1912. Armstrong, whose favorite book was *The Researches, Writings, and Inventions of Nikola Tesla,* said of his Columbia professor, "[Pupin] went so far as to say there was very little originality about Tesla."[15]

In the case of Westinghouse, after Tesla's patents lapsed, the Tesla motor and AC polyphase system came to be known as the Westinghouse motor and system. As I explain in my Tesla biography, *Wizard,* even at the Westinghouse Company, there were many key individuals who did their best to obscure Tesla's role in the massive success of the company. Frankly, had it not been for John O'Neill, a reporter for the *Herald Tribune* who personally knew the inventor for forty years and whose biography on Tesla, *Prodigal Genius,* was published in 1944, Tesla's name would have almost certainly completely disappeared from the history books. As unbelievable as it sounds, for decades, all the way through the 1970's, the only real way to obtain the O'Neill biography was through a UFO organization who distributed the book. The lesson here is the importance of the individual, in Tesla's case, in making history, and in O'Neill's case, of preserving it.

All of this may seem tangential to a discussion of Tesla's life story because his name has reappeared in a major way in books, movies, and on the Internet. But it is important for serious researchers to realize that ego battles, personality conflicts, and specifically, corporate and other financial interests play a key role in obscuring history and, ultimately, in this instance, thwarting the development of technology. Not only was radio and cell phone technology retarded by decades, the release of fluorescent lightbulbs was also delayed nearly a half-century because corporate powers that were backed by J. P. Morgan simply wanted society to spend more money in illuminating their homes and factories with incandescent lights that expend most of their energy in heat, not light, and that burn out after just a few months.

Tesla invented cold fluorescent bulbs in 1898, yet they did not appear on the market until the 1940s, and that was because large buildings simply had to use them because the demand for electricity had grown so rapidly. In fact, we have finally reached a stage whereby the Edison lightbulb may be phased out completely, to be replaced by Tesla fluorescent lights, which, for the last half-century, have been the dominant way factories and buildings are lit. Also, where an Edison incandescent bulb may last a few months, a Tesla fluorescent bulb can last literally decades.

ACCOLADES AWARDED TO TESLA

In 1893, Tesla was awarded the Elliott Cresson Gold Medal from the Franklin Institute in Philadelphia. This was followed by the Edison Gold Medal given to Tesla in 1917 and the John Scott Medal Award in 1934. Tesla was also awarded advanced degrees and honorary doctorates from Columbia University and Yale University in the United States, and from universities in Vienna, Bucharest, Zagreb, Belgrade, Prague, and Paris. In 1937 he was nominated for a Nobel Prize, in 1976 a statue of Tesla was erected on Goat Island at Niagara Falls on the U.S. side in Tesla's honor, and in 2006, another statue of Tesla was erected by the Horseshoe Falls on the Canadian side.

As the twentieth century ended, in 1999 *Life* magazine listed Nikola Tesla as one of the one hundred most important individuals of the last one thousand years. His picture can be found on the equivalent of the one-dollar bill in Serbia and there is a U.S. stamp that commemorates him. Tesla was one of our greatest visionaries, very far in advance of his time. He remains as an inspiration to the übergeeks of Silicon Valley today. A tremendous resurgence of interest in Tesla has occurred because of the Internet, a global communication system very much in accord with his inventions and predictions.

> *The scientific man does not aim at an immediate result.*
> *He does not expect that his advanced ideas will be readily*
> *taken up. His work is like that of the planter for the future.*
> *His duty is to lay foundation for those who are to come,*
> *and point the way. He lives and labors and hopes.*
>
>   NIKOLA TESLA, 1900

2

JOHN WORRELL KEELY

FREE-ENERGY PIONEER—A NEW CHAPTER

Theo Paijmans

> *Let a note be struck on an instrument, and the faintest*
> *sound produces an eternal echo. A disturbance is created on*
> *the invisible waves of the shoreless ocean of space, and the*
> *vibration is never wholly lost. Its energy being once carried*
> *from the world of matter into the immaterial world will*
> *live forever.*
>
> H.-P. BLAVATSKY

> *My system, in every part and detail, both in the*
> *development of this power and in every branch of its*
> *utilization, is based and founded on sympathetic vibration.*
>
> JOHN WORRELL KEELY

INTRODUCTION

He was the source of an endless series of controversies printed in newspapers worldwide during his career that spanned a quarter of a century.
Many tens of thousands of dollars were invested in his inventions. He

Figure 2.1. John Worrell Keely surrounded by his inventions

won over the hearts and minds of even the most sceptical, cynical businessmen with spectacular demonstrations. He built hundreds, some say two thousand, of the strangest engines the world has ever seen. He used a phraseology and terminology few could understand. He claimed to have found a limitless source of energy locked up in the grids of the atoms with which he could generate antigravity and disintegrate matter. He was the subject of a ruthless exposé after his death. His secret—if he ever possessed one—was lost forever with his demise. All of his constructions save one were lost forever too. There are no personal papers or accounts of his research in existence, and it is doubtful that he actually wrote or assembled such papers at all.

All these aspects form but a part of what is still a complex enigma, more than a century after his death. The name of this man was John Worrell Keely, and even today opinions on him are sharply divided. Charlatan or genius, avant-garde philosopher or obfuscating swindler? It is a question that haunts us still. In studying his exploits, each corner in his kaleidoscopic career holds a new surprise. There is so much that is contradictory that the mind is constantly swayed toward both ends of this spectrum by each turn we take in the study of his life and work. For instance, if he was a charlatan, why then were all the funds he procured apparently invested in his researches and the construction of his engines? And if he was a mountebank, why does the only device that is with us today—an incomplete construction—still elicit admiration for the skill, refinement, and incredible workmanship that went into creating this apparatus? On the other hand, if Keely was truthful, why was he unable to produce a working engine in the twenty-five or more years he was involved in his researches, notwithstanding his many promises to do so?

THE EXTRAORDINARY STORY OF
JOHN WORRELL KEELY.

Figure 2.2. Keely and his famous Dynasphere

*Figure 2.3. Keely and friend with yet another of Keely's
marvelous inventions*

So who was Keely? Was he a self-made scientist operating on the absolute fringes of science, and did he discover by chance something so extraordinary that he was unable, or was not allowed, to replicate it for the rest of his life, as some have suggested? Or did he himself exert an influence over his devices, an influence that he himself may not have fully realized, one belonging more to the realm of the paranormal, as others have theorized?

This is the enigma that is John Worrell Keely. Having written the only biography of him, titled *Free Energy Pioneer: John Worrell Keely,* published in 1998 on the centennial of his death and still in print, even I cannot answer all these questions with a ready response. Having no vested interest either way, my research then as now focuses primarily on reconstructing an accurate historical picture of Keely and his discoveries, something that was clearly lacking when my biography appeared. Keely, I quickly found out when I started my research in the early 1990s, was already the stuff of legend, of the folklore that clings to the borders of the history of Western technology and invention, with precious little in the way of primary and original sources. I was able to remedy this situation with my biography and correct many an error that had crept into the fierce debate on what Keely had been and what he accomplished. Until invited to write this essay, I never suspected to return to this topic—one I treated so exhaustively during the five years it took me in writing my book.

The debate on Keely between his apologists and detractors rages on, now as it did then, and I doubt we will soon see the end of it. This paper introduces Keely once again in the larger history of the search for free and alternative energy sources, with the benefit of introducing new and pertinent materials not known or available to me at the time of the publication of my biography. This prompted me to write this paper, so those already familiar with my book and with Keely have new materials at their disposal. What this also means is that the mystery of who Keely was, and precisely what influence his ideas have had in other fields, is still a valid subject for research, with the chance always of uncovering new venues to explore. To those unfamiliar with Keely's life and work I offer a detailed overview of one of the strangest careers in the nineteenth century. And perhaps one day, one of the readers of this paper may stumble on the key that finally unlocks the secrets Keely took to his grave: access to that fabulous force locked between the atoms of the ether, and thus usher in a new golden age for mankind.

KEELY'S LIFE

Biographical information on Keely is fragmentary, sketchy, and sometimes conflicting. He was born September 3, 1837, either in Philadelphia in a two-story frame house that stood on the corner of Jacoby and Cherry streets or in the old town of Chester, Pennsylvania. Sometimes his birth year is given as 1827. Keely was twice married, the second time in 1887. He left no children. His first wife and only child had died many years before. Of his second wife, only her name, Anna Keely, is known. Keely died in Philadelphia on November 18, 1898. He was buried in the West Laurel Hill Cemetery in Philadelphia, where he has plot number 313 in the River Section. His grave was unmarked for a century, a situation remedied by Dale Pond and others, who placed a gravestone on the centennial anniversary of his death on November 18, 1998.

His mother's parents were of English and Swedish descent, his father's of German and French descent. His father was an ironworker.

Keely lost his parents in his infancy. His mother never recovered from his birth, and his father died before he was three years old. Keely had one brother, J. A. Keely. Since Keely was orphaned in early childhood, they were probably separated early. After his father died, Keely came to live with his grandmother and an aunt, but his aunt passed away before he was sixteen, followed by his grandmother a year later. By then, Keely had already left school by the age of twelve.

Somewhere in these early childhood years Keely made his astonishing discovery and decided to follow that path for the rest of his life. Although contemporary sources contradict each other in certain details, the consensus is that Keely had his moment of illumination while observing peculiar effects of sound on certain objects. Keely too claimed that he started his researches in his childhood, leading him to his discovery that would haunt him for the rest of his life: "Before I had reached my tenth year, researching in the realm of acoustic physics had a perfect fascination for me; my whole organism seemed attuned as if it were a harp of a thousand strings; set for the reception of all the conditions associated with sound force as a controlling medium, positive and negative; and with an intensity of enjoyment not to be described."[1]

Drawn as he was to this line of research at such an early age, an incident several years later set him on the course that he decided to follow: "The first manifestation . . . when I was twelve years old drew my attention to the channels in which I have since worked."[2] He never told what that manifestation, incident, or discovery exactly was.

In his pre–motor fame days Keely held a number of jobs to sustain himself. Contemporary accounts have it that he was employed as a physician, a pharmacist, an upholsterer or a cabinetmaker, a plumber, a plater, and a mason. One contemporary source observed that during all this, "He had an inventive genius and gave much more attention to mechanical problems than to his employment."[3]

On only very rare moments Keely himself confided details of his early life to a reporter, and we perceive a glimpse of what must have been a most unusual career at times:

> Are you a spiritualist, Mr. Keely? I asked him. That is one of many
> lies propagated about me, he answered. It has been said I started life
> as a carpenter (though that is not a slander), but I didn't: I never
> was a carpenter. Instead of being a spiritualist myself, I once exposed
> spiritualistic mediums in St. Paul, Minn., in 1857, 1859, and 1861, and
> I was nearly run out of town for doing so. Everything their mediums
> did in the dark, I did in the light, and that naturally enraged them. I
> do not believe in Spiritualism or in anything of the kind. I am, I hope,
> a Christian, and a regular member of the Methodist Church. . . .[4]

There are claims that Keely actually created a number of curious devices or prototypes of engines very early on in his career. We have a description of what may be one of Keely's very first devices, but unfortunately no date or time period is given. This device "for noting the uniform force of sound vibrations" consisted of a steel bar set full of pins of various lengths, while his first "resonator" or "intensifier" consisted of a shingle screwed to two hollow wooden tubes. Keely's first rudimentary engine was also described as "a simple ring of steel with 300 pins set into it, and this first wheel ran in an open box, into and through which an observer was free to look while the wheel was in motion."[5]

In 1856 Keely was experimenting with a toy engine, the boiler of which was fired by a "burning fluid" lamp in his home on Fifth Street, Philadelphia. In 1863, Keely became employed as a furniture varnisher in the furniture shop of a certain Bennet C. Wilson. Wilson financed Keely's experiments and provided and furnished a place on Market Street in which he installed Keely. There Keely experimented "year after year" with an engine that he called a "reacting vibratory motor." This led in 1869 to the construction of a device that can be considered the forerunner of his globe motor.

Around 1874 Keely moved to a building at 1420 North Twentieth Street in Philadelphia, where he established his workshop. It was a modest dwelling that had been formerly used as a stable. Keely used the first floor as a general storeroom, where he also conducted his experiments. The second story consisted of three apartments, the first being the office of the inventor, the second the workshop for his globe motor then in development, with an adjoining room. He would occupy this workshop till his death. Currently, the location where his laboratory once stood is a parking lot. Nothing reminds one of the incredible life and career of one of the most enigmatic figures in the annals of free and alternative energy research. But here he would conduct his experiments, hold many demonstrations, and astonish reporters and witnesses with glimpses of an utterly strange technology, such as a large device meant as an engine for an airship. But the practical applications of this technology, as always, were still far away. This remained so till the day he died.

Keely passed away on November 18, 1898, after having contracted a severe cold. His funeral services were held on November 23, and a large crowd turned out. Anna Keely placed a red heart near the coffin, one half to be buried with Keely, the other half to be taken home with her.

During his life a Keely Motor Company was formed in order to fund his work. Stock worth tens of thousands of dollars was sold, but in the end the stockholders were left empty-handed as well. In his incredible career, Keely interacted with a host of remarkable historic figures. Thomas Edison and Nikola Tesla turned down invitations to witness demonstrations of his engine. John Jacob Astor invested in him. The founder of theosophy, Helena Petrovna Blavatsky, wrote about him in her *The Secret Doctrine,* and after her, a host of other theosophists did so too.

Moreover, he met with a remarkable person who would not only become his most staunch and trusted advocate, but owing to whom many of the connections above came about. She wrote many pamphlets and one book about Keely, and also aided financially to sustain his incredible career. This was Mrs. Bloomfield-Moore, maiden name Clara Sophia Jessup (1824–1899). She supported Keely from 1882 till the end, something that is still slightly frowned upon by close family members even today. Yet Bloomfield-Moore was an extraordinary and highly intelligent woman who developed a deep interest in matters scientific and occult. She corresponded with hundreds of scientists in all parts of the world, including Tesla, whom she met at least once, writer on the fourth dimension Charles Howard Hinton, and early aviation pioneer Hiram Maxim. At one time she offered financial support to Maxim if he would go to America to consult Keely and "become the custodian of the latter's secret."

Around Bloomfield-Moore, we see a clustering of some of the most famous, wealthy, and influential persons of her day, such as her wealthy acquaintance John Jacob Astor, who once was courted by Nikola Tesla to fund his research. She almost constantly remained in London after the death of her husband in 1878. Bloomfield-Moore was also presented to the court of Queen Victoria. Both her daughters married nobility. One daughter married Swedish Baron Carl von Bildt, who at one time had been secretary of the Swedish Legation in Washington. Her other daughter, Ella, married Count Carl von Rosen, who was first chamberlain and master of ceremonies at the Court of Stockholm, Sweden. It was a son from this marriage, Count Eric von Rosen, who, as it was later claimed, allegedly sent unspecified materials pertaining to Keely—named "Keely's secrets"—to Sweden.

But there was also another side to her complex character. She gained a mild reputation with the public for eccentricity, possibly because her

Figure 2.4. Count Eric von Rosen (1879–1948) as he was in 1919. It is said that he sent "Keely's secrets" to Stockholm in 1912. Von Rosen also was brother-in-law to Hermann Göring.

London home witnessed visits of those of the occult circles. Bloomfield-Moore's writings in themselves are remarkable for their deep esoteric knowledge, but she was not a theosophist herself, according to Count von Rosen, when he was asked this obvious question: "No. She was interested in the study of theosophy as a broad-minded woman. She was interested speculatively, but did not believe in it."[6]

Although not a theosophist, she developed a relationship with Blavatsky that is described as "long" and "intimate," and Blavatsky published excerpts of her writings on Keely in *The Secret Doctrine*. As a rather interesting detail, but not one without consequences as we shall see, a son of the marriage of her daughter Ella, Count Eric von Rosen, had founded the Finnish Air Force in 1918 by presenting the White Army its first air-plane, a Thulin Parasol fighter aircraft. Count von Rosen, a noted archae-ologist, had his arms painted on both wings' upper and lower surfaces of the aircraft presented. His arms were a blue swastika on a white back-ground. He had seen these on rune stones in Gotland while still at school.

Bloomfield-Moore died only a few months after Keely, and her literary executor, science writer Henry Dam, said, "I knew that when Mr. Keely died she would not live long. Her whole life was centered in his work to the exclusion of all other interests and hopes. She had the most profound faith that neither Mr. Keely or herself could die until the invention had succeeded. After receiving the cabled announcement of Keely's death she began to sink rapidly. Her ailment seemed more mental than physical."[7]

A devastating exposé followed, initiated by Clarence Moore, who had always opposed his mother's backing of Keely. After Keely died, Moore and a number of scientists entered his workshop, where, as published in the newspapers, they found hidden tubes, wires, and a motor in the base-ment. Their verdict was simple: there was no mysterious force, and there never had been. All these years, Moore claimed, Keely had bamboozled his audiences by spinning his engines with the help of these hidden tubes through which compressed air was pumped, and with hidden levers, springs, and that machine in the basement.

KEELY'S DISCOVERIES

Since we have no detailed records by the hand of Keely himself, try-ing to establish a coherent pattern in his researches and inventions is a daunting task. Although there is an abundance of contemporary sources, with recollections of persons who each gave their viewpoints and offered various statements on the matter, we still are faced with a jumble of dates, descriptions, and recollections sometimes made years after their alleged occurrences and the fact that Keely suddenly and radically altered his line of research in the 1880s.

Roughly speaking then, there were two periods in Keely's researches and experiments. The first period ran more or less from 1874, the year he started to occupy his workshop, to 1882. This period may very well have started some years earlier, considering the various tales of prototype devices he had constructed as early as 1856 and possibly even before.

It is also alleged that Keely's motor began to attract attention as early as 1865.[8] Around 1871, Keely, again working to putting his discovery to practical, commercial use, announced his invention to the public. Keely, or so it is said, placed an ad in a Philadelphia evening paper "relating to a new motor, or motive power, which he alleged he had invented or discovered and was prepared to exhibit."[9] This new motor was named the globe motor, presumably because it consisted of a hollow sphere that revolved at great speed and, as Keely declared, automatically. This first period was marked by the production of force, what was later called etheric vapor, by the disintegration of water. This Keely accomplished with several of his engines, such as a device known as his liberator, and we have several accounts that serve as an illustration of Keely's experimental activities during that period. It was explained that the etheric vapor was obtained by letting the water into the disintegrator or liberator at a certain pressure. Its aim was to disintegrate water into etheric vapor.

In 1871 Keely pursued his investigations in an effort to work out this discovery, using the two elements water and air in connection with sound vibrations. In 1872 Keely allegedly made his discovery of an energy that he called "the force" by accident while experimenting with vibrations. He then "imprisoned the ether" the same year and "commenced his experiments with ether in the winter on 1872–'73."[10] At first Keely had no idea what he had found, as he readily admitted. It would take him twelve years before he realized he had "imprisoned the ether."

Between 1871 and 1875, Keely also constructed six different devices. At that time thirty-four documents were in existence relating to the transfer of interests in inventions that were called an independent flywheel, the hydro pneumatic pulsating vacuo engine, the globe motor, the dissipating engine, the multiplicator or generator, and the automatic water lift. The first assignment was dated July 11, 1871, the last, February 15, 1875.[11]

Keely's globe motor, although exhibited in operation around 1871, was never patented. What was found during an investigation in 1875 was "on record in Liber L, 18, page 370, of transfer of patents, U.S. Patent Office, an assignment of this so-called globe motor by Mr. Collier to the

Keely Motor Company, this assignment bearing date February 15, 1875, and being recorded May 8, 1875."[12] The patent office at that time also had an abstract of "all assignments, agreements, licenses, powers of attorney and other instruments in writing on record" in the name of John W. Keely since January 1, 1871. However, there does exist a patent by Keely, granted August 15, 1871, for his fly-wheel, an arrangement of gearing for causing a wheel to revolve at a greater speed than the shaft to which it is hung.[13]

In 1872 Keely constructed a new motor at his place in Market Street. This second motor was variously called the hydraulic motor, the hydro vacuo engine, the hydro pneumatic vacuum engine, and the hydro pneumatic pulsating vacuum engine. A New Yorker who claimed to have known him during that time later wrote, "I was with him when the idea first entered his head that he could combine steam and water to run an engine. At that time he made a crude machine, which he actually ran for some time; and this was the model of the Pneumatic-Pulsating-Vacuo-Engine. . . . In those days I have known him to sell and pawn everything of value in his house to obtain means to continue his investigation with the money thus acquired."[14]

This model was subsequently located at 1010 Ogden Street in Philadelphia, where Keely was then living as well. The model was described as an engine placed in a bathtub and run by a stream of water that passed through a goose quill. This device "soon grew into the machine which he called a 'generator,' and which the world named the Keely motor, and in which power was produced from the vibratory qualities of water and air."[15] Elsewhere it is claimed, though, that the generator not so much evolved out of his hydro pneumatic pulsating vacuo engine, but that after its construction, Keely "took a new departure," which culminated in the so-called Keely motor, or, as it has been termed, "a dissipating engine and multiplicator and generator."[16]

While working with his generator one day in 1873, Keely suddenly felt a cold vapor blow in his face. He tried to wipe away the moisture but was surprised to find "there was none upon his countenance. The curious phenomenon of a vapor that was absolutely dry caused him to take up a new line of experiment."[17] This mysterious vapor was described as "a heretofore unknown gaseous or vaporic substance," and it was the power on which the generator—also termed the dissipator or the Keely motor—worked.[18]

> [Keely being] . . . a poor man, but, having a wonderful degree of natural mechanical skill . . . devoted all his time for the past fourteen years to experiments with water with a view of procuring a motive power from it. He was engaged upon an idea of his own regarding the force of columns of water one day when he accidentally discovered the vapor which he has harnessed. He studied the subject, ascertained how it was generated, learned its power, and thenceforth applied himself solely to the perfection of this idea, working night and day for a number of years, until his efforts were crowned with success.[19]

Since the above quote is taken from an article written in 1875, this would imply that Keely was involved in this line of research since 1861. Elsewhere it is claimed that in 1873 Keely became known as the discoverer of a new power, "which he had not then been able to utilize, to operate machinery, but which could be supplied in limitless quantities at practically no cost. . . . He said himself that he made the discovery in 1872, but then had no idea of its origin or laws. He gave no indication of its character, but kept the secret within his own breast until such time as patents could be secured."[20]

Keely did attempt to secure a patent on his device. There exists a patent application by Keely that was filed on November 14, 1872, titled "Specification describing a new and useful Hydro Vacuo Engine, invented by John W. Keely of the City and County of Philadelphia and state of Pennsylvania." Keely's machine is described as, "The end and design of the invention is an engine wherein the actuating power is produced by a vacuum in connection with water pressure."[21]

Yet nothing came of his application. When information on this application was gathered around 1875, a search of the records of the patent office brought to light an abandoned application for a patent for the hydro-vacuo engine.

. . . the said application having been filed November 14, 1872. . . . At the request of the applicant's attorney, a model was dispensed with by the authorities in the first instance; but on November 26, 1872, a working model was demanded before the examination could be completed. Whenever an application for a patent is of doubtful practicability, or based on what is believed to be a fallacy, it is the practice of the patent office authorities to demand a working model, and to refuse to examine the case until the demand is complied with. Nothing was done in this case of Keely's until March 20, 1874, when he appointed Mr. J. Snowden Bell, now the mechanical associate of Mr. Collier, to prosecute the application; but as two years elapsed without any action, the application was thereby under the law abandoned.[22]

There are rumors of subsequent patents and the assignment of rights. Wilson for instance claimed that Keely had assigned to him "one full half ownership" of the principles or machine that he built in 1869. The other rumors stem in all probability from the 1871 patent, which was granted directly to the assignees, two unidentified persons who at first advanced money for the further development of the globe motor and the hydro-pneumatic-vacuo engine, or his patent application in 1872.

A mysterious patent that Keely supposedly requested on November 26, 1873, and of which it was remarked that the accompanying drawings "are now lost," never existed. In contemporary sources there is no further reference to other patents.

The machines that Keely built during his first period were of megalithic proportions, one weighing as much as twenty-two tons. Most were implemented by the Atlantic Works and the Delaware Iron Works of Philadelphia. They were scrapped in turn as they were superseded by smaller, more sophisticated models. His generator of 1878 weighed about three tons and occupied a space five feet long and high with a width of two feet. It contained small spherical chambers, "mathematically differentiated in size," connected vertically by "tubular processes of iron, and irregularly by smaller ones of copper. One quart of water fills all the chambers and tubes intended to be filled."[23]

The generator, upright in position, had five distinct parts or columns, called the central column, two side columns, and the front and back stand-tubes. The stand-tubes, although similar in appearance, were opposite in action. The two side columns were alike. The central or main column was larger than the other four combined and was "more complex in structure." Air was water-locked in some of the chambers and tubes, where "introductory impulses" were fed to the water so that its equilibrium was disturbed. This disturbance was effected by the movement of an outside lever operating a four-way valve within. There were no other metallic movements inside except the working of three independent valves. The apparatus was therefore considered "practically without wear and not liable to get out of order."[24]

The generator used but one quart of water to produce fifty-four thousand pounds of pressure per square inch. No heat, electricity, or chemicals were used, and it was claimed that output remained constant regardless of the work effected. The generator was one of two mechanisms Keely built during this period, the other being the engine. The generator produced the force that the engine used. These two devices were what is commonly referred to as the Keely motor. The vaporic substance was the medium of the force that it carried.

The second period in Keely's researches ran from 1882 until Keely died. Keely had for ten years made demonstrations of the liberation of the energy he claimed to have discovered while experimenting with vibrations in 1872. But his efforts to construct an engine that he had promised the Keely Motor Company failed. The explanation on offer is that Keely was pursuing his researches on the line of invention instead of discovery, and all his thoughts were concentrated in this direction up to 1882. Frequently, in his constant failures to construct an engine that would keep up the rotary motion of the ether, explosions occurred, and Keely scrapped engine prototypes and sold their remains as old metal.

In this period of constant frustration, Keely discovered a force derived from the vibration of an unknown fluid or substance, locked between the atoms of the ether. This he called the vibratory force or the etheric force. It was this force that, according to Keely, was not like steam, electricity, compressed air, or galvanism. While, during the first period, Keely obtained his force through the disintegration of water, during the second period Keely developed his force "in the air, in a vacuum, in the ether itself," or by the vibrating of hydrogen, which Keely first attempted around 1884 on the suggestion of Bloomfield-Moore.

Up until 1888, Keely continued trying to construct a "perfect engine," which could hold the ether in "a rotating ring of etheric force." Toward the end of 1888, Keely entirely abandoned his concept of the perfect machine. Up to this time he practically built his research equipment himself. Around 1888, Keely was to be provided with the best instruments opticians could build for him based on the models or designs he furnished.

During this second period the miniaturization of his engines continued for a considerable degree. From 1882 to 1884 the Generator was a structure six feet long and correspondingly wide and high. A machine built in 1885 that he named the Liberator was somewhat smaller than "a lady's round worktable." Continuing on this path of miniaturization, Keely "within one year made such an astonishing progress . . . as to combine the production of the power, and the operation of the cannon, his engine and his Disintegrator in a machine no larger than a dinner plate, and only three or four inches in thickness. This instrument was completed in 1886."[25] When Keely began experimenting with his discovery of another principle—one of the reasons the Keely Motor Company took him to court, the others his failure to secure patents on his former discovery and his failure to construct a commercially succesful device—his engines became even smaller in size, "and the size of the instrument used now, '88, for the same purposes is no larger than an old-fashioned silver watch, such as we see in Museum collections."[26]

In his productive career Keely built between 129 and two thousand experimental devices, but as far as is known, never once more than one of each. This enormous number was reached because of the special requirements that Keely had. His engines needed a perfect construction, and if a device possessed but one little unevenness or imbalance, he considered the engine worthless, one of the reasons for this being that Keely worked with enormous pressure forces. In his search for perfection, vast sums of money were expended on machines that sometimes would not even be used when ready. One device had cost $40,000, and when ready was rejected by

Keely because he claimed there was "a flaw in it." Keely "thought nothing of spending $20,000 for a piece of machinery, and a few weeks afterwards throwing it aside as useless."[27]

Other devices were made of special metal alloys; his generator was made of Austrian gun metal "in one solid piece." It would hold about ten or twelve gallons of water and was four or five inches thick, made "to handle the very heavy pressure of 20,000 to 30,000 pounds of vapor to the square inch." Other parts were made of welded iron "of great thickness and strength." Then there was the adjusting of the machinery, which seems to have been a difficult affair "when it is considered that Keely experimented with 129 machines during his career as an inventor, and that one after another of these was made to special measurements, to be thrown into the scrap-pile. . . . Many of these machines called for the most delicate adjustment and the most consumate skill of the artisan."[28]

Figure 2.5. This broken glass plate was discovered by Jeff Behary, hidden away for almost a century in Kinraide's basement laboratory. As far as we know, this is the last photo made of Keely's globe motor. Photo © Jeff Behary, c/o the Turn of the Century Electrotherapy Museum

Figure 2.6. Keely and colleagues at the height of his most inventive stage

Of all the engines Keely built, photographic evidence exists of only a mere fraction. The devices on these photos have such wonderful names as the compound disintegrator, the vibratory globe machine, the resonator, the rotating globe, which worked through human magnetism, a number of vibratory discs, the spirophone, the pneumatic gun, the provisional engine, the globe motor, the vibrating planetary globe, the wave plates, the planetary system engine, the liberator, the vibrodyne or vibratory accumulator, the vibratory switch, and the sympathetic negative transmitter. It is the same with written descriptions of his devices. For these we have to rely largely on eyewitness accounts scattered in contemporary newspapers.

At the time of Keely's death in 1898, a new engine was in its completion stage, which Keely expected to have in running order the following year. The machine had the same shape as Keely's globe motor, only larger. It was made out of copper with a globe about two feet in diameter and weighing seventy-five pounds. The final mechanism was three feet in diameter and built of decarbonized steel, weighing six hundred pounds. Its whereabouts—if it still exists—are currently unknown.

NEW DEVELOPMENTS

More than half a century after Keely's death and the ensuing tumult over his legacy and his devastating exposé, Keely's nephew was located

JOHN WORRELL KEELY 61

and interviewed. He was an old man at that time and professed that he never had much confidence in Keely's experiments, but he recalled how Bloomfield-Moore sent "many of Keely's secrets" to Count Eric von Rosen. He apparently sent this material to Stockholm in 1912, and nothing more was heard of these "secrets." Perhaps these consisted of the materials, mainly photographs of various engine-like devices with the handwriting of Bloomfield-Moore on several of them, which were published in 1972 in Sweden.

Eric von Rosen was also to become brother-in-law to Hermann Göring when the sister of his wife married Göring. Göring, the World War I ace aviator, once flew Eric von Rosen in bad weather from Stockholm to Rockelstad Castle. Due to this bad weather Göring had to stay over. And while he noted a swastika over the hearth place, it was his meeting with his future wife, as she descended the stairs in that ancient castle, which connected him with a very secretive sisterhood, the Edelweiss Gesellschaft. There is a letter in existence where Göring describes his feelings, seated in contemplation in the small chapel that this sisterhood had.

Although this strange turn of events would have no place in the history of Keely, it is well worth mentioning it for the following reason. In seeking to establish what quarters upheld an interest in Keely's discoveries even after his exposé, ironically conducted by Bloomfield-Moore's son Clarence, I found an intriguing allusion pertaining to certain experiments conducted in the Third Reich. Briefly, the context of these experiments was the search for alternative means of creating fuel for the German war machine, as Germany had no oil reserves of its own. But it did have large quantities of coal. It used a process called the Fischer-Tropsch method to create oil from gas. But there is a hint that there was another process in existence with which Nazi Germany was allegedly able to refine oil into various usable components using certain sound waves. Or so it was claimed by prisoner of war Josef Ernst to his English captors.

The document in question, the British Intelligence Objectives Subcommittee Report Number 142, reads in part, "Ernst claimed to have invented a process for the separation of petrol from oil by the use of vibrations of audible frequency. This method was developed from the Kelly process in use in the U.S.A. The crude oil is passed through glass tubes suspended above a plate of 'pertinax,' which is caused to vibrate at frequencies on the Pythagoras tone-scale. The frequency is altered according to the product required. . . . This process was operated at the wells near Hamburg owned by the S.S. and many tons of petrol were produced. These factories were probably destroyed, but the apparatus is possibly in the south of Germany."[29] It is clear from the description that by "Kelly" John Keely is meant.

So on one hand we have a direct connection between one of the future leaders of the infamous Third Reich and the one person alleged to have transported certain materials pertaining to Keely to Sweden.

While there is no verifiable documentation at hand that would enable us to reconstruct without doubt the conduit through which information on Keely traveled to the heart of the dark empire, if at all, we must bear in mind that news on Keely was published in nineteenth-century Germany as well. These coincidences give food for thought and may merit a closer inspection into the irrationalist belief systems of Nazi Germany, especially in regards to the many rumors about new and outré forms of technology said to have been in development there. Be this as it may, the statement above by German prisoner of war Josef Ernst is, to my mind, the only contemporary account of a technology either in use or in development based on Keely's discoveries, even long after he had been discarded by conventional science and he was delegated to the occult fringes associated with theosophy. And even there the exposé by Clarence Moore caused some uncomfortable mutterings.

Then there is the subsequent fate of some of Keely's engines, interwoven with the strange story of Thomas Burton Kinraide of Jamaica Plain, Boston. The first direct reference to Kinraide—apart from appearing on the list of pallbearers at Keely's funeral—appeared in the press eight days after Keely's death.

> [Kinraide] . . . came on to Philadelphia after Mr. Keely died, was at the house every day between the death and the funeral; rode with Mrs. Keely in the same carriage to the cemetery and returned with her to the house. . . . Several years ago, it is said, he appeared at the Keely

Figure 2.7. Thomas Burton Kinraide (1864–1927), last inheritor of Keely's legacy. Photo © Jeff Behary, c/o the Turn of the Century Electrotherapy Museum

establishment and has been a constant visitor and intimate of the family ever since. He remained on some of his visits as long as four or five weeks, during which time he devoted day after day to the study of the Keely motor, remaining in the room with it for hours at a time.[30]

A contemporary investigation into the background of this man yielded some intriguing details:

It was discovered that he was a man of great wealth and lived in princely fashion in the vicinity of Boston. It was also learned that he was a scientist with theories of his own almost as remarkable as those advanced by Keely. Like Keely, he pursued discovery on the tones of vibration, and in order to conduct some special experiments, he had a huge cave hewn out of the solid rock on his estate, which he fitted up as a laboratory. During experiments, Kinraide had succeeded in obtaining rotary motion on the compass needle from vibrations.[31]

On December 28, 1898, Keely's engines were shipped to Boston. But, contrary to what has always been suggested by later researchers into this part of the mystery, not all of Keely's engines were to go to Kinraide's laboratory:

All the devices of the late John W. Keely that are held to be of importance in connection with the experiments to be made by T. B. Kinraide, of Boston, will be shipped to-day, to be placed in his laboratory. The more delicate machines and parts have been in safe keeping in the vaults of the Land Title and Trust Company, having been placed there shortly after the death of Mr. Keely. Charles S. Hill, of Boston, attorney for Mrs. Keely, was in the city yesterday, and with Mr. Kinraide supervised the work of moving the machines. . . . Mr. Hill says that the older machines and those which Mr. Kinraide believes will be of no practical value in his work, will be stored by the Keely Motor Company in this city, and that a room will be rented for that purpose.[32]

Hill personally escorted Keely's engines to Boston. Whatever was left of Keely's engines at that time still must have been substantial since its inventory, filed on January 3, the same day Keely's devices arrived in Boston, made a reference to "fifteen pieces of experimental apparatus," and a Boston newspaper wrote, "Twenty large packing cases have arrived here, containing the material part of the famous Keely motor. T. B. Kinraide, an inventor, ordered the boxes removed to his laboratory. There he will experiment in trying to supply whatever is lacking in mechanism."[33]

Interviewed by a reporter, Kinraide told him that

he had often talked with Mr. Keely on the principles of his invention. He never fully explained the secret of his perpetual motion to me, said Mr. Kinraide, but I feel that I know more of the motor than any other man. Mr. Keely, after being taken ill, expressed the wish that I be allowed to carry out his inventions. Before the hour set for the interview had arrived the inventor was past recovery. It was, however, at Mr. Keely's request and that of Mrs. Keely that I have consented to conduct these experiments.[34]

So Keely's engines, which were already crated on December 20, were at this stage stored at two places: the contents of twenty large crates were on Kinraide's estate on Jamaica Plain in the vicinity of Boston, where his laboratory was situated, and in Philadelphia, where Keely's "mechanical property" was removed and stored in a building on North Broad Street. Apart from the machinery that was moved to Boston, the "heavier apparatus" was relocated to a storehouse at Broad and Vine Streets, possibly the same storehouse on Broad Street. In a safe deposit vault of the Land Title and Trust Company were placed Keely's disintegrator and other "fine pieces of machinery," his sensitized disks, wires and other objects. Before being placed in the vault, the disintegrator made a short stop at the Hotel Stratford, where it was examined by John J. Smith of the Keely Motor Company, Kinraide, and Hill.

Perhaps a detailed inventory was made of the devices left at Keely's death and showing which devices Kinraide obtained, but if so, it has not survived. We are therefore left with superficial descriptions that Kinraide obtained Keely's "remarkable machines, vibrators, lever machine and others used by Keely." It was suggested that Kinraide also obtained Keely's latest engine, work on which was almost finished at the time of his death. It resembled Keely's globe motor but was larger, with a two-foot copper globe weighing seventy-five pounds and the mechanism made out of decarbonized steel weighing six hundred pounds.

Then Clarence Moore once again made a detrimental statement. He told a reporter that

Mr. T. B. Kinraide, of Jamaica Plain, Boston, on whom the mantle of Keely was supposed to have fallen, and who actually did receive the Keely motor mechanism early in January, admits that the motor was a fraud, that the machinery was moved by well-known forces, and that the duplication of Keely's "demonstrations" is a simple matter. When Mr. Kinraide was reported, a short time ago, to have returned the Keely motor machinery to Philadelphia, the inevitable conclusion was forced on any one conversant with the history of the Keely motor that Mr. Kinraide, having discovered the nature of the fraud, washed his hands of the whole affair. It seems however, that Mr. Kinraide did not send the apparatus back, but got to the heart of the delusion by a careful study of the motors in hand.[35]

According to an unnamed informant, during this "careful study" Kinraide had fitted up the machine room in his cave at Jamaica Plain "almost exactly after the manner of Keely's, suggesting that he had begun on the premise that he must reproduce the whole thing if desirious of success."[36] Then, Kinraide had walked around the machines for some time, and "at last he turned and smilingly remarked that he might excel Keely at his own trickery if he had the same fluency of words as the latter."[37] Kinraide had applied compressed air, hydraulic pressure, and a powerful spring, and even a magnet concealed in the wall. The hydraulic pressure and compressed air came from "hidden sources." With the help of all this, Kinraide obtained the same results. "One of the most unique pieces of mechanism I found in Philadelphia," Kinraide reportedly said, "was a spring, to wind which it is necessary to use a key as big as a crowbar." With the proper winding the spring would be able to "run for three or four days" and produces enormous energy. Kinraide also showed how he could start the hydraulic and compressed air pressure by picking up a violin, after which the "instrument wheels began to revolve" because he touched "a bulb" hidden under the floor at the same time.[38]

It took the unnamed informant nearly eight hours spent in Kinraide's laboratory, "and when the Philadelphian emerged he was convinced that he had lost his investment."[39] Kinraide repeated that those who would produce evidence of being "victimized" could obtain an invitation for visiting Kinraide's laboratory and "see the whole thing disclosed." Kinraide refused entrance to reporters, "for it has been decreed that, as far as the public is concerned, the Keely affair can rest in peace and the Keely victims have their sorrows to themselves."[40] So goes the Clarence Moore report.

Since publishing my research results in 1998, new information with a startling outcome in regard to Thomas Burton Kinraide has come to light. These new discoveries were made by Jeff Behary, owner and proprietor of The Turn of the Century Electrotherapy Museum in Florida. I am happy to publish, in his own words, his quest for Thomas Burton Kinraide, his trek to his onetime residence at Jamaica Plain, and the stunning treasure that he found hidden away, untouched for decades. The following words are Behary's.

JEFF BEHARY'S ACCOUNT OF THE HIDDEN TREASURE

My research of Thomas Burton Kinraide began around 1995. I had begun reading a fascinating book called *High-Frequency Currents* by Frederick Finch Strong. This book was unusual on many levels—its primary subject was the therapeutic use of Tesla coils. This book is filled with rare information, but one paragraph stood out in my mind for no clear or apparent reason—a paragraph where Strong describes witnessing X rays of enormous power in Kinraide's laboratory. He stated that Kinraide was the first person to X-ray the entire human trunk, and then went on to describe the apparatus. Strong was a curious man on many levels. He invented a form of vacuum electrode that later flooded the medical Tesla coil market and received no credit for it in time. A decade later, he decided to write the history of Tesla coils and give himself his proper spot in the timeline as it were. Had he not done this, the information would be lost today, not only of his work but his friends as well as competitors. It was an unusual thing to do, because the field itself was relatively new—he wrote the book in 1908, and most of the subject matter was post-1896. But, as often is the case, the first decade of an invention's existence is often the most important.

Perhaps it was a realization that this great man Strong was first inspired by this mysterious Kinraide fellow. I sought to find more on who Kinraide was, because on the pages that followed I saw a photo of Kinraide's machine and the description was quite unusual as far as Tesla coils were concerned. It appeared on many levels to be an engineering marvel of its own right, and even the appearance was quite unlike the hundreds of similar apparatus that flooded the market later on. This apparatus was unique, unusual, and by Strong's description, efficient beyond imagination.

For years I searched with little success. I was relentless, sometimes searching until 2:00 a.m. in vain for information. It became an unusual obsession, an almost impossible quest. Finally, I decided to try and recreate Kinraide's apparatus only from the brief description in Strong's book. To my surprise, I had instant success with this unusual venture. I immediately posted photos to my website, mainly in efforts to share the marvelous efficiency of my crude machine to my friends and colleagues. In the Tesla arena, this was virtually uncharted territory—and that is always exciting. These were the days of dial-up modems, and in updating my site I would confirm the sending of everything before bed and simply check everything upon waking up the next morning. It sometimes took hours.

I knew Kinraide was from Boston, in a small town, Jamaica Plain. That night, I dreamed that Rita and I traveled to Boston in search of Kinraide's work. In my dream, we were on a bus and there were two laughing and giggling teenage girls sitting ahead of us. My wife and I were discussing the idea of opening a coffee shop (which we had been discussing in real life). The girls turned around, and one looked me dead in the eye and said, "If you like coffee, go to the June Bug!" I immediately awoke from the weird dream and stumbled into the other room to check if the uploads worked. They were complete. It was around 2:30 a.m., and I decided to Google "June Bug Boston." To my surprise, there was a coffee shop with the name "June Bug Café" *in Jamaica Plain*. How unusual! I couldn't believe it.

The next day, I told my wife of the story, who rolled her eyes a bit at yet another story of me and my Kinraide quest. I never forgot that dream. I refined my Kinraide invention and slowly incorporated other of his inventions into its construction. The next reproduction of Kinraide's invention was around a year later, and that night I didn't have any strange dreams. But the next day, I received a call out of the blue from Gerald Zeitlin, a retired anesthesiologist and historian, who introduced himself as standing by the grave of William James Morton in Forrest Hills Cemetery, *Jamaica*

Plain, Massachusetts. I was astonished. Morton was a friend of Strong and Kinraide's. I asked Gerald if he knew the June Bug, and he laughed, remarking that it was a small shop down the road. I instantly booked tickets to Boston. If the dream wasn't a strange form of "sign," the phone call definitely was. To wait a month for this travel was agonizing, so I focused on building a small Kinraide invention to pass the time. The next day, I found a Kinraide patent for sale in England. I bought it immediately and had it shipped airmail, international express. I had only two weeks and feared I wouldn't receive it in time. However, against all odds, it arrived in a few days. Written on the patent was Kinraide's street address in Jamaica Plain. I had not expected this, because in U.S. patents only the county was required information. I Googled the address and found a phone number. The poor owner must have been bewildered to receive a call from a complete stranger regarding the man who built his home. "Kinraide built this house, but is a bit of a legend here. No one knew who he was."

Figure 2.8. Ravenscroft: the house of Kinraide, with a large laboratory in the basement where Keely's engines were shipped for experimentation. © Jeff Behary, c/o the Turn of the Century Electrotherapy Museum

He was kind enough to grant me permission to visit his home. I was not expecting the mansion that awaited me. Having said that, days before I called him just to confirm when I could meet him. Unfortunately he was called out of town for a family emergency, but he still granted me access to his beautiful house through the generous help of one of his employees. Once in the house, I was shown around twenty-five rooms, but the most unusual was a hidden ballroom—a secret door to which was a moving wall next to a large safe. The walls were all solid mahogany, and there were hunted taxidermy animals that Kinraide shot that looked partially alive, as if they were watching the inhabitants of the home. It was surreal. There were beautiful brass chandeliers that could be used as gas lights if the power failed. The home was elegant beyond imagination. I asked to see the basement. I was taken downstairs with Rita, and in the basement I noticed a blocked door that appeared as a small closet. I asked to see inside, and the man showing us around remarked if we wanted in there I'd have to move the host of items that were blocking the entrance myself. I did so gladly, and on opening the door noticed a stack of old windows and some wood. As I turned around in this room, which was nearly black from the lack of lighting, I saw one wall appeared to go a bit farther than the outside would hint to. It was actually much larger, and using the flash of my camera for eyes, I inched my way to what seemed to be another room. It was! I saw an old radio, a large painted board that read "Twinkle Twinkle Little Star," and stacks of mahogany. Then there appeared to be crates of some kind. Still blind with only the camera flash for sight, I snapped photos and saw that the crates contained glass negatives. I was almost shaking as I removed one from the crate. The plates were of sparks, and I knew that Kinraide made electrographs that, according to a top X-ray pioneer, "would have astonished Faraday beyond words."

Moments later I picked up a glass negative and saw the Keely motor. I was almost in tears. My wife, who for years had put up with my madness in this pointless quest, *was* in tears when I showed her the photo. More than a decade of illogical searching was suddenly worthwhile. We managed to find an extension cord and had a single lamp. The rooms were blasted out of rock like tunnels in caves, and they continued to circle around the basement. The next room contained the remains of Kinraide's experiments that I read about in 1995, ten years prior. The room that followed contained more mysteries and treasures. I asked the man who was helping us to call the owner of the home. I explained to him that I found hidden treasures beneath the house, and this complete stranger told me that my search was meant to be and to take them! It was approaching evening, and I asked if we could return the following day (Saturday) to pack up the items. The man agreed, and we went to dinner. It was the most exciting day of our lives.

However, that night at dinner, Rita remarked that I lost my wedding ring. I looked down and couldn't believe it. The ring was normally so tight on my finger I could only remove it after minutes of prying with soap and cursing at the abrasions of twisting and turning it over my knuckle. The ring was platinum, and my wife spent quite a bit of money to obtain it. We couldn't forget it amongst the other treasures. So the next day, the man let us back in the basement and told us he had to go and would be back in a few hours. My wife took a stroll to the local post office to get a heap of priority mail boxes. I was alone in the house. In looking through some of the negatives, I noticed in the dim lighting of the room that there was handwriting on the moldy paper envelope that they were contained in. I decided to walk outside to have a better look in the light.

I took out the camera and decided to document a historic find of Kinraide's handwriting. However, after the flash, I looked down to see the part of the word I was trying to decipher was gone. Perhaps the fragile paper fell apart when I was walking? I retraced my steps carefully, and it was nowhere to be found. I was bewildered and frustrated at this; what's the chance of finding an autograph and then *losing* it! As I went back down to the basement, a large oil furnace that was completely silent previously started to rumble and thunder and make a lot of noise. I heard that they could be noisy, but didn't imagine in this way. I then got a feeling as if I was being watched. Not the typical feeling though; this time every hair on the back of my neck stood up.

I was actually afraid. I turned to look at the entrance of the rooms and sensed something or someone. I took a photo. I looked at the photo. At that point I froze in place and was shaking. I never saw such a shape in the tens of thousands of photos I've taken before. It was as if something was sitting in the doorway. I wondered if I was correct in my actions at the home. My wife returned and found me in a state of fright. I only handed her the camera, and on seeing the image she was amazed!

The picture matched an electrical form that Kinraide called an "electric entity." I also later learned it matched so-called orb photos, as well as dust particles captured by flash photography. However, under the circumstances, the premonition of fear leads me to believe that T. B. Kinraide wanted to acknowledge my presence. And then, typical of my wife, she remarked, "I don't care if there's a ghost. You're still finding your wedding ring!" The first thought was to check where the coils were, as I had to dig a bit in the dark through wax and rosin to find them. We had no luck. But knowing of Kinraide's failed marriage, Rita exclaimed while digging, "I know your marriage wasn't good, but it doesn't mean you can take our ring!" I took her picture.

We never found the ring. We did pack thirty-two boxes and managed to ship them from the local post office. They were happy to see us leave after that! Thirty of the thirty-two boxes arrived on Monday back in South Florida. Next to the stack of boxes, at around 3:30 p.m., I walked up to the front door and saw my wedding ring on the ground. Perhaps it fell out of one of them. Or maybe, just maybe, Kinraide decided to give it back!

Since then I returned to Ravenscroft once. I searched with Kinraide's granddaughter's husband for signs of the Keely motor. We had little time, and the frost on the ground made it hard to decipher what was under the frozen dirt/rock/rotting leaves floor. We did find the original oil-filled X-ray coil that Kinraide made the first intense X rays with, as mentioned in the Strong book. Kinraide placed it in an oak box lined with copper. A curious chain assembly was also found inside this box, and after further inspection we found that he had been running the coil in the back of a Victorian toilet tank! I guess if it was watertight, it would be oil-tight too! Plus he could "flush" the oil to change connections and fine-tune the coils before reinsulating with oil. History is stranger than fiction.

In walking out, I found in the area where the original Keely negatives were a single piece of torn paper with the words "Keely M . . ." written on top. Part of the original envelope that was rotting off of the negative I found two years earlier. I believe that Kinraide demonstrated the motor in the rooms I was exploring. I personally believe that there were elements of fraud in Keely's motor. Either way, I think it makes him one of the most interesting people in American history. If he was a fraud, so be it, but a genius, well, there was no question. I am sure that any negative publicity that came out of this brought Kinraide much grief on many levels. Who would want to see a friend defamed, much less be the one to have to deliver the news? Since the discovery of Kinraide's lab, I have learned countless things that should have been lost a century ago. The Keely motor is only one piece of the puzzle. In time, it might be surprising what we find. If I hit the lottery for millions, I would offer every cent to buy that home. I would give up millions, even at the thought of remaining virtually penniless, just to pursue the dream that wasn't pointless!

And so, in closing with Jeff Behary's remarkable account, we have taken the history of the Keely enigma one step further. We were able to uncover cryptic traces of what appears to have been a technological process in Nazi Germany based on Keely's discoveries, and we have been able to travel inside Kinraide's laboratory—the final resting place of Keely's mysterious engine. There is still much to be uncovered, and who knows what the future has in store for us?

3

VIKTOR SCHAUBERGER

A BRIEF OVERVIEW OF HIS THEORIES ON ENERGY, MOTION, AND WATER

Callum Coats

How could we have missed this universal machine? Why have we ignored the vortex, the workhorse of the universe?

WILLIAM BAUMGARTNER

VIKTOR SCHAUBERGER, THE MAN

Throughout recorded history humanity has been periodically uplifted by the contributions of a few gifted and enlightened individuals whose teachings and philosophies have gradually raised the level of human awareness, with Buddha, Jesus Christ, and the Prophet Mohammed being the most illustrious examples of individuals who have produced far-reaching changes in the consciousness of humanity. Lesser mortals have also played a vital role in this process, and the seeding of human consciousness with higher truths always seems to come at a time when humankind as a whole is ready to receive them.

It is sometimes said these great teachers, themselves ardent students of Nature and the Divine, live ahead of their time. These exceptional individuals are indeed visionaries in the truest sense of the word, for they are endowed with a far higher sense of perception than their con-

*Figure 3.1. Viktor
Schauberger, Austrian
Forester and Natural
Scientist*

temporaries. For their work an enormous dedication and courage is necessary. Historically—and Viktor Schauberger was no exception—the lives such individuals have led, or have had to lead, have been dogged with confrontation, difficulty, doubt, and the great loneliness of the pathfinder. As pioneers, apart from breaking new ground, they often suffer great adversity in their encounter with the powerful opposition of those whose interests and beliefs are rigidly immured in the current status quo.

Schauberger's life followed this same path, for in his life he too was met with derision, slander, and deceit in a long confrontation with the establishment in its various forms. More often than not his discoveries contradicted established theory and in their flawless functioning and practical implementation seriously threatened the credibility and reputation of scientist and bureaucrat alike. Schauberger was a man of enormous strength of purpose, warm and encouraging, particularly to young people, in whom he took great interest, for he saw in them the possibility of restoring a secure and bountiful future. But to those whose view of life he considered irretrievably perverted, spiritually and intellectually, he was absolutely uncompromising, seeing them as obstacles in the path of human evolution and the rehabilitation of the environment.

There are many such individuals who have given themselves wholly to the betterment of their fellow human beings. Without exception they were endowed with extraordinary perceptive and intuitive abilities that afforded them fresh insights into the way the world functions, enabling them to understand phenomena hitherto inexplicable to their contemporaries. They were aware of another dimension of reality, one further dimension at least always being required to make sense of the whole. Analogous to the third dimension that makes a two-dimensional world understandable, this can be called the "dimension of comprehension."

Schauberger, some of whose penetrating insights into natural phenomena we will address here, was one of those rare human beings, those explorers in human thought and endeavor whose chosen path was to throw light on the future. In years to come he will be acknowledged as one of the principal guiding spirits of the twenty-first century and beyond, who brought about a fundamental shift of Copernican proportions in humankind's appreciation of Nature and natural energies. Indeed, few have had Schauberger's deep understanding of that living substance so vital to all life processes—water, which he viewed as the blood of Mother Earth, for Schauberger saw the whole Earth as a living organism.

Schauberger was born June 30, 1885, in the parish of Ulrichsberg, Upper Austria. He was descended from a long line of foresters who had devoted their lives to the natural management and administration of the forest—a dedication mirrored in their family motto: *Fidus in silvis silentibus* ("Faith in the silent forests"). With this as his background, and much against his father's will, at the age of eighteen he flatly refused to follow in the footsteps of his two elder brothers and attend university, having seen how university had affected his brothers' thinking. The main reason for his refusal was that he did not wish to have his natural way of thinking corrupted by people he considered totally alienated from Nature. He did not want to be forced to see things through jaundiced eyes, but through his own. As he later wrote, "The only possible outcome of the purely categorizing *compart*-mentality, thrust upon us at school, is the loss of our creativity. People are losing their individuality, their ability to see things as they really are and thereby their connection with Nature. They are fast approaching a state of equilibrium impossible in Nature, which must force them into a total economic collapse, for no stable system of equilibrium exists. Therefore the principles upon which our actions are founded are invalid, because they operate within parameters that do not exist."[1]

Endowed with an exceptional gift for accurate and intuitive observation, Schauberger was able to perceive the natural energies and other phenomena occurring in Nature, presently unrecognized by orthodox science. Leaving home, Schauberger spent a long period alone in the high, remote forest, contemplating, pondering, and observing the many subtle energetic processes taking place in Nature's laboratory, where still undisturbed by human hands. During this period he developed very profound and radical theories, later to be confirmed practically, concerning water, the energies inherent in it, and its desired natural form of motion. These eventually earned him the name Water Wizard.

While Schauberger undoubtedly had a special talent for observation, a penetrating power of perception unsullied by preconceptions, he also developed what might be called an active consciousness, an ability to go beyond the merely visual in search of what lay behind a given phenomenon. This taught him a great deal, and how this ability gradually evolved he explained as follows:

The Schaubergers' principal preoccupation was directed towards the conservation of the forest and wild game, and even in earliest youth my fondest desire was to understand Nature and through such understanding to come closer to the truth; a truth that I was unable to discover either at school or in church.

In this quest I was thus drawn time and time again up into the forest. I could sit for hours on end and watch the water flowing by without ever becoming tired or bored. At the time I was still unaware that in water the greatest secret lay hidden. Nor did I know that water was the carrier of life or the ur-source* of what we call consciousness. Without any preconceptions, I simply let my gaze fall on the water as it flowed past. It was only years later that I came to realize that running water attracts our consciousnesses like a magnet and draws a small part of it along in its wake. It is a force that can act so powerfully that one temporarily loses one's consciousness and involuntarily falls asleep.

As time passed I began to play a game with water's secret powers; I surrendered my so-called free consciousness and allowed the water to take possession of it for a while. Little by little this game turned into a profoundly earnest endeavor, because I realized that one could detach one's own consciousness from the body and attach it to that of the water.

When my own consciousness was eventually returned to me, then the water's most deeply concealed psyche often revealed the most extraordinary things to me. As a result of this investigation a researcher was born, who could dispatch his consciousness on a voyage of discovery, as it were. In this way I was able to experience

---

*In Schauberger's writings in German, the prefix *Ur* is often separated from the rest of the word by a hyphen (e.g., *Ur-sache* in lieu of *Ursache*, when normally it would be joined). By this he intends to place a particular emphasis on the prefix, thus endowing it with a more profound meaning than the merely superficial. This prefix belongs not only to the German language, but in former times also to the English, a usage that has now lapsed. According to the *Oxford English Dictionary*, *ur* denotes "primitive," "original," and "earliest," giving such examples as "ur-Shakespeare" or "ur-origin." This begins to get to the root of Schauberger's use of it and the deeper significance he placed on it. If one expands on the interpretation given in the *Oxford English Dictionary*, then the concepts of "primordial," "primeval," "primal," "fundamental," "elementary," and "of first principle" come to mind, which further encompass such meanings as pertaining to the first age of the world or of anything ancient; pertaining to or existing from the earliest beginnings; constituting the earliest beginning or starting point, from which something else is derived, developed, or depends; applying to parts or structures in their earliest or rudimentary stage; and the first or earliest formed in the course of growth. To this can be added the concept of an "ur-condition" or "ur-state" of extremely high potential or potency, a latent evolutionary ripeness, which given the correct impulse can unloose all of Nature's innate creative forces.

things that had escaped other people's notice, because they were unaware that a human being is able to send forth his free conscious-ness into those places the eyes cannot see.

By practicing this blind-folded vision, I eventually developed a bond with mysterious Nature, whose essential being I then slowly learnt to perceive and understand.[2]

These perceptions of truth presented Schauberger with considerable problems in translating them into everyday language, for when it comes to transferring spiritual ideation into mundane word-pictures, enor-mous difficulties are encountered due to the limitations of language. In many instances, when he came to describe these phenomena, he described them not in the conventional terminology of physics, chemis-try, or biology, but in his own words, for which he was greatly assisted by the structure of the German language, which facilitates the forma-tion of new concepts through additive nouns. Despite this, and for lack of a suitable technical vocabulary, their interpretation and comprehen-sion is still sometimes extremely difficult, which in his writings he freely admitted: "Few will understand the meaning of the above! Some indi-viduals, however, will obtain an indefinable inkling."[3] His oft-repeated dictum was "C^2 —Comprehend and Copy Nature," for only thus will humanity emerge from its present crisis-stricken condition.

The vital development of a new technology, harmonious and con-forming to Nature's laws, demands a radical and fundamental change in our way of thinking and our approach to the interpretation of the estab-lished doctrines and facts of physics, chemistry, agriculture, forestry, and water management. As a pointer as to how such a new technology should come about, let me again quote Schauberger: "How else should it be done then?" was always the immediate question. The answer is simple: *exactly in the opposite way that it is done today!*

However, before we can usefully address Schauberger's theories, it is first necessary to discuss energy and movement for, contrary to the generally accepted way of looking at things, in his view of life's pro-cesses and their unfoldment, *energy is primary and physical form is the secondary effect.*

WHAT IS ENERGY?

What is the essential nature of energy? Where do we begin to search for the answer to this age-old, deeply philosophical question? Surprisingly, despite all scientific investigation, nobody seems to have come up with a definitive answer! All we know are the ways in which energy manifests itself. We can see that energy is involved in flowing water and the formation and movement of clouds. But what is energy? What is its essence? What is this sublime process that always seems intimately connected with motion?

There are many extremely high energies of which science is aware and can measure, but it cannot measure human energies such as thought, desire, love, enthusiasm, anger, and such, all of which are expressions of the human psyche and motivators to action. Many ancient cultures have known of other immaterial life energies, which they called, for example, chi, ka, prana, mana, archeus, or vis vitalis. As Schauberger often said, scientific thinking is an octave too low and is unaware of what he called the fourth and fifth dimensions. As human beings we are immersed in a three-dimensional world, yet have an inkling of a possible fourth dimension we call time. What other dimensions may await our discovery?

The Creative Energy Vortex

It is becoming more imperative that we understand how energy moves in order to create conditions in any future technology that faithfully emulate the natural movement of energy and Nature's systems of motion, growth, and development. In her systems involving dynamic energetic processes, she always appears to select a spiral form of movement and its vortical derivatives, which are represented in figure 3.2 in both macrocosm, a galaxy, in this case overlaid by Walter Schauberger's hyperbolic spiral, and microcosm, a DNA molecule.[4]

Since we still do not know what energy is, for the purposes of discussion, figure 3.3 represents a possible energy path. As energy moves along its desired path, it draws matter into its wake and forms the vessel through which it wants to move. A river does exactly the same thing; the capillaries in our bodies likewise. The blood is the external manifestation of an energy path. What we see is the blood; we do not see the energy that moves it. The blood is all that matter that is too coarse to

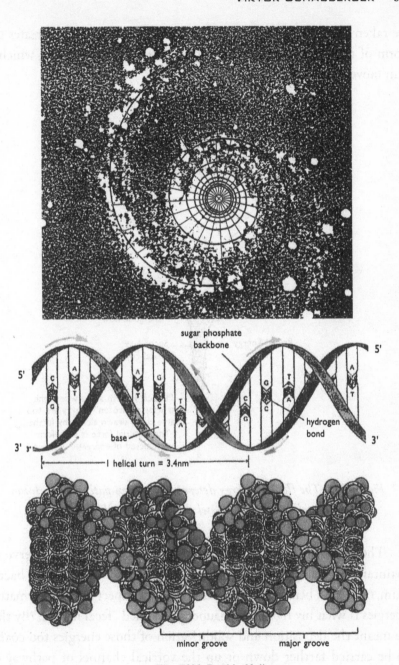

Figure 3.2. *In forming systems of dynamic energy, Nature always seems to select a spiral form of movement, from the incredibly huge whirlpool of galaxies to the incredibly small spirals of DNA.*

be taken to the energy's final destination. Energy therefore creates the form of the path through which it wants to move and along which it can move with the least resistance.

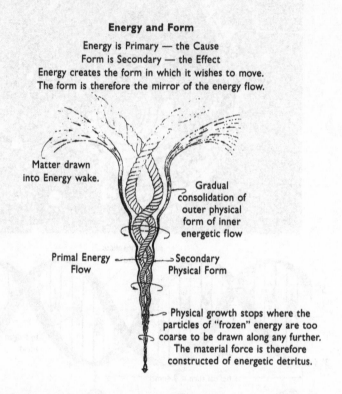

Energy and Form
Energy is Primary — the Cause
Form is Secondary — the Effect
Energy creates the form in which it wishes to move.
The form is therefore the mirror of the energy flow.

Matter drawn
into Energy wake.

Gradual
consolidation of
outer physical
form of inner
energetic flow

Primal Energy
Flow

Secondary
Physical Form

Physical growth stops where the
particles of "frozen" energy are too
coarse to be drawn along any further.
The material force is therefore
constructed of energetic detritus.

Figure 3.3. The flow of energy determines its own path, and it draws matter to itself to form the vessel through which the energy flows.

These unseen energies do not cease there, however, but serve to maintain the energy of the organism in question, be it a tree, bacterium, or human being. The physically palpable aspect of these formative energies is what my mentor, Schauberger, termed "fecal matter." By this he meant the deposition and solidification of those energies too coarse to be carried farther down or up the vortical channel or pathway of the continuously vivifying and form-maintaining energies. From a certain point of view, the human body could therefore be seen as a hollow energy path, a complex toroidal vortex for the transmutation of matter-energy into physical and intellectual activity.

In Schauberger's view the unseen vortical energies responsible for the growth and development of the tree, for example, stream upward far above the physical tree itself and draw up its structure in their wake, the additional growth resulting from the accumulation and expulsion of further energetic detritus.

The use of the word *vortical* in reference to the pathways of these formative energies relates to the fact that vortices have a cohering, energizing function. In their implosive outside-inward motion, they draw the various energetic filaments together into closer harmonious association and thus maintain a concentration in their concerted action to create the artifact in question. Depending on their frequency, there may be several such vortices emanating from the original vortical flow or life-stream, each being responsible for the creation of a different aspect of the evolving form.

In our physical world vortices and spirals are evident in both material and immaterial form, and in their physical manifestation these energy vortices display the most exquisite mathematical perfection and symmetry, of which seashells, flowers, and leaves are perhaps the most familiar examples. In terms of physical growth this motion can only be outward, as in the case of the seashells, for physical growth cannot occur in an inward direction, but only outwardly. Immaterial growth, however, can take place in either direction, but in vortical systems, such as cyclones and those in water, the movement is from the outside inward, coupled with increasing density and velocity. While the implosive vortex is the predominant one, Nature is never one-sided, and therefore the opposite form exists and is implemented when required.

Of their very nature these vortical systems are self-energizing, friction and impedance reducing, self-organizing, spatially reducing, cooling, suction increasing, densifying and cohering, and silent. It is a form of motion that Schauberger termed radial $\rightarrow$ axial (i.e., movement from the outside inward). In contrast, all our current technical systems operate in the opposite direction (i.e., inside outward, or axially $\rightarrow$ radially). Such motion is coupled with increasing pressure, friction, disintegration, and ultimately noise. Indeed it could be said that the more noise a technical system makes, the more it is operating against the laws of Nature. Such noise, however, has a debilitating and disruptive, if not downright harmful, effect on all natural organisms forced to endure it, including us.

Continuing the above analogy of the vortical movement of energy, let us observe just how beautiful such a naturally structured vortex is (figure 3.4). Such phenomena are not often observed. What a marvelous structure! It is not handmade, but it is the path along which water likes to move.

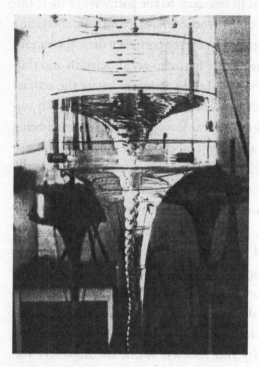

Figure 3.4. A photograph of the energy path of a naturally structured vortex

Nature's workings could therefore be described not as "wheels within wheels" but as "whorls within whorls." It is all the more extraordinary, therefore, that despite so much evidence of this vortical, cyclical, and helical movement, which lies everywhere in Nature before our very eyes, science has never ascribed any fundamental importance to it or tried to copy it. It has been too immersed in the euclidean elements of mechanics with little knowledge or conception of organics. We have never taken the time to understand it enough to be able to exploit it. It is high time we developed a technology whereby these processes are truly understood. This should be termed an ecotechnology rather than a biotechnology, the latter having been brought into disrepute through gene manipulation and experimentation. Perhaps "ec²otechnology" would be an even better term as it embodies Schauberger's concept of C^2, signifying "Comprehend and Copy Nature."

The "Original" Motion

If one observes the universe as a whole, from big bang to black hole, as it were, a form of motion is evident that Schauberger called "cycloid-spiral-space-curve motion." He also referred to it as the "original" motion, not only in a primordial sense, but also as a "form-creating" dynamic. Shown in its quintessential, archetypal form in figure 3.5, which depicts the creation of three successive universes, the cycloid-spiral-space-curve embodies an initial out-breathing, centrifugal, curving expansion from a point, which results in the generation of countless individualities and energetic systems. Its culmination is an in-breathing, centripetal implosion whereby all that has been created is concentrated once more into a point. The very word *universe* signifies a single curve (*uni* = one, *versum* = curve). The fact that the configuration of this curve may be a complex combination of descending and ascending, involuting and convoluting, expanding and contracting spiral movements does nothing to detract from its uniqueness or unit quality, since from inception to culmination its path is continuous. It is an energy path, and the essence of energy is ceaseless movement. In its eternal trajectory from spirit to matter (outward breath) and from matter to spirit (inward breath), it permeates all creation. It is all creation!

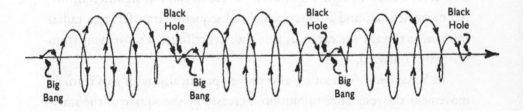

Figure 3.5. Creative, formative motion according to Schauberger. The open, goal-oriented, structured, concentrated, intensifying, condensing, dynamic, self-organizing, self-divesting of the less valuable, rhythmical (cyclical), sinuous, pulsing, in-rolling centripetal (and out-rolling centrifugal) movement equals the cycloid-spiral-space-curve.

Apart from its inherent pulsation, it would be impossible to break this eternal movement down into discrete segments, for at the point where one portion of this sublime curve ceases, the next begins and cannot be defined mathematically, whatever the subjective view. Therefore this unique, primordial, creative curve embodies the unbroken path of evolution, of cyclical, pulsating out-foldment and in-foldment, as it spirals in and out of all the myriads of apparently inextricably interconnected and interdependent individual systems in the cosmos, tying and uniting all in one inscrutable Gordian knot.

Even the tools of common language unwittingly allude to the character of this creative force and its dynamic spiral movement. When we "ex-spire," we exit from this our "mortal coil." When we are "in-spire-d," we feel drawn to higher ideals. Our "spir(e)it" is raised, and we are sucked into the upward spiral.

Interestingly enough, the German word for the spinal column, the fundamental supporting structure of the human body, is *Wirbelsäule,* which translated directly into English means "spiral column." Similarly each one of the vertebrae is referred to as a whirlpool or a vortex. Whereas we see it as a stiff, more or less rigid, physical structure, the Germans see it more as an energy path.

Forms of Motion

When we come to spiral-vortical motion itself—and Nature provides us with countless examples—we can further subdivide it into another two forms. As shown in figure 3.6, axial → radial motion signifies an initial movement around a center, which subsequently transfers to a radial movement toward the exterior; it is thus centrifugal, a movement from the inside outward.

In Schauberger's theories, also proven practically, with this form of movement the resistance to motion increases by the square of the starting velocity. In other words, if the radial distance from the center of rotation is 1 and the resistance is 1, when the radius is doubled (from 1 to 2), the resistance is (2 squared = 4) quadrupled and the rotational period halved. If the radial distance is 3, the resultant resistance is 3^2 (9) and the rotational velocity reduced to one-third, and so on. However, if the rotational velocity of such a centrifugal system is to be maintained at a constant level, then a continual, wasteful, and expensive increase in the amount of input energy is required to overcome the resistance, and the whole system becomes less and less efficient. Not only this, but it creates discordant noise, and the more noise a device makes, the more it operates against the laws of Nature.

The dispersion of energy, therefore, is associated with noise or heat, as the case may be. This is typical of our forms of technical movement, wherein there is initially no motion at the center, but with increasing distance from this point, velocity and resistance also increase. The axial → radial centrifugal form of motion can thus be defined as divergent, decelerating, dissipating, structure loosening, disintegrating, destructive, and friction inducing. While the destructive diffusion of energy results in noise, the creative concentration of energy, however, is silent. Indeed, as Schauberger asserted on many occasions, "Everything that is natural is silent, simple and cheap."[5]

Upon consideration, this statement is quite obvious. Think of all the concentrated energy involved in the growth of the forest, for example, in all the innumerable chemical and atomic interactions, which are none other than energetic processes, movements of creative energy. The silence of the forest is indicative of the extraordinary concentration of form-creating energy.

Whereas our mechanical, technological systems of motion almost without exception are axial → radial and heat and friction inducing, Nature uses precisely the opposite form of movement. When Nature is moving dynamically, the slowest movement occurs at the periphery and the fastest at the center. One only has to observe the dynamics of a cyclone or a tornado. Her form of movement, therefore, is centripetal or radial → axial, moving from the outside inward with increasing velocity, which acts to cool, to condense, and to structure.

Radial → axial motion can therefore be defined as convergent, contracting, consolidating, creative, integrating, formative, and friction reducing. If the starting radius is 1 and the initial resistance is 1 on an inwinding path, when the radius is halved, the resistance is $(\frac{1}{2})^2 (= \frac{1}{4})$, and the rotational periodicity, frequency, or velocity is doubled. The dynamics of evolution must therefore follow this centripetal, radial → axial path, for if the opposite were the case, namely centrifugal, axial →radial motion, then all would have come to a stop almost before it started. With centrifugal acceleration, more power must be applied in order to accelerate or to maintain the same velocity. However, if the acceleration is centripetal, the velocity and energy increase automatically, producing a creative force, the upbuilding energies from which all life is created.

As we are all becoming increasingly aware, our axial → radial technology and all its unnatural appurtenances, encompassing such effects

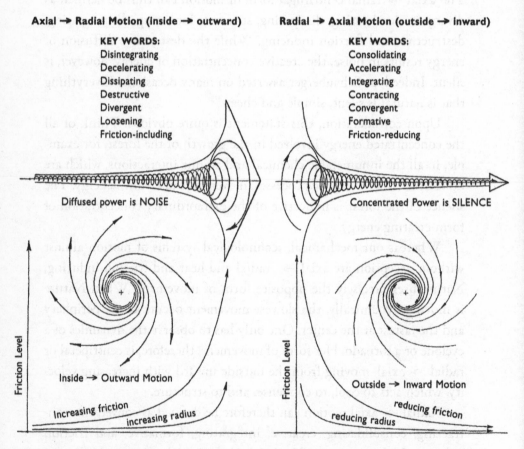

Axial → Radial Motion (inside → outward)

KEY WORDS:
Disintegrating
Decelerating
Dissipating
Destructive
Divergent
Loosening
Friction-including

Diffused power is NOISE

Friction Level

Inside → Outward Motion

Increasing friction
increasing radius

Radial → Axial Motion (outside → inward)

KEY WORDS:
Consolidating
Accelerating
Integrating
Contracting
Convergent
Formative
Friction-reducing

Concentrated Power is SILENCE

Friction Level

Outside → Inward Motion

reducing friction
reducing radius

Figure 3.6. Schauberger's concept of axial → radial motion denotes a movement from the center (the axis) outward toward the exterior of a vortex (the radius). Such movement is therefore centrifugal (i.e., from the inside outward).

as pollution, toxic waste, monoculture, and artificial fertilization, to name a few, is reaching the point where it is threatening our very existence, creating a crisis of life and death.

The Chinese word for crisis encompasses the two elements of "danger" and "open door." Thus when danger confronts, a door opens to avoid it. This is where we are right now, and in preparation many alternate and Nature-friendly technologies are being quietly developed and can be implemented. Such implementation, however, will require a volte-face in humanity's attitudes, not only toward each other, but also in our interactions with Nature. Fundamental to any new and harmonious culture, however, is a reevaluation and a reverence for water, whose natural movement embodies all the forms of motion described above and without which there would be no life at all. Thus it is to water that we shall now turn our attention.

WATER—A LIVING SUBSTANCE

Water! Where do we begin our quest in search of the true nature of this remarkable substance, this wondrous, many-faceted jewel, which is both life and liquid? So primordial, primeval, and fundamental is the function of water that it begs the question: Which came first, life or water? Thales of Miletus (640–546 BC) described water as the only true element from which all other bodies are created, believing it to be the original substance of the cosmos. It was the only real substance because it was imbued with the quality of being.

This view was firmly held by Schauberger, who also saw water as the "original" and "form-originating" substance created by the subtle energies called into being through the "original" motion of the Earth, itself the manifestation of even more sublime forces. Because it is the offspring or the "first born" of these energies, as he put it, he maintained and frequently asserted that *water is a living substance!*

Because he saw water as a living entity, Schauberger also saw it as the accumulator and transformer of the energies originating from the Earth and the cosmos and, as such, as the foundation of all life processes and the major contributor to the conditions that make life possible. Not only that, he said, but once mature, water is a being invested with the power of extraordinary giving and gives of itself to all things requiring life.

> *The Upholder of the Cycles which supports the whole of Life, is water. In every drop of water dwells a Deity, whom we all serve; there also dwells Life, the Soul of the "First" substance—water—whose boundaries and banks are the capillaries that guide it and in which it circulates.*
>
> VIKTOR SCHAUBERGER,
> *OUR SENSELESS TOIL*

Water is therefore a being that has life and death. With incorrect, ignorant handling, however, it becomes diseased, imparting this condition to all other organisms, vegetable, animal, and human alike, causing their eventual physical decay and death, and in the case of human beings, their moral, mental, and spiritual deterioration as well. From this it can be seen just how vital it is that water should be handled and stored in such a way as to avert such pernicious repercussions. Rather than the nurturer and furtherer of all life that it should be, failure to perceive water as a living entity quickly transforms it into a dangerous enemy, for when, in our ignorance of water's manifold functions, we stop these cycles, we also stop life.

As a liquid, water is described chemically as H_2O and is a dipole molecule composed of two hydrogen atoms, each endowed with a positive charge, and one oxygen atom possessing two negative charges. Water is no homogeneous substance, however, for it possesses other characteristics according to the medium or the organism in which it resides and moves. As a molecule, water has an extraordinary capacity to combine with more elements and compounds than any other molecule and in this regard could be described as the universal solvent. At a physical level water is to be found in three states of aggregation, solid (ice), liquid (water), and gas (water vapor), and in terms of its structure as a liquid it tends more to the crystalline, as it continually forms and reforms nodes of temporary crystallization.

The Properties of Water

Anomaly Point

The anomalous expansion of water is also a factor of major importance. As a liquid, the behavior of water differs from all other fluids. While the latter become consistently and steadily denser with cooling, water alone, among all liquids, reaches its densest state at a temperature of +4° Celsius (39.2° Fahrenheit), below which it once more expands, eventually crystallizing as ice at 0°C. The temperature of +4°C is the so-called anomaly point or "point of anomalous expansion," which is decisive in terms of its potency and has a major influence on its *quality*. At +4°C water attains its greatest density, its least spatial volume, and is virtually incompressible (figure 3.7).

If water's temperature rises above +4°C, it expands. When it cools

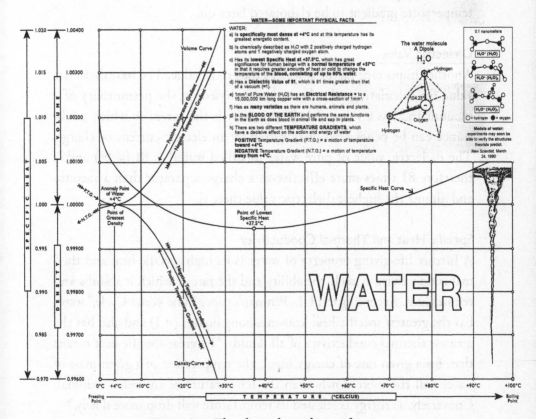

Figure 3.7. The physical attributes of water

below this level it also begins to expand and becomes specifically lighter. This anomalous expansion below +4°C is vital to the survival of fish life, for as water expands and cools further it eventually crystallizes as ice at 0°C, providing a floating, insulating sheath that protects the aquatic life underneath from the harmful effects of severe external cold in winter.

Also, +4°C is the temperature at which water has its highest energy content and is in what Schauberger called a state of "indifference." In other words, when in its highest natural condition of health, vitality, and life-giving potential, water is at an internal state of energetic equilibrium and in a thermally and spatially neutral condition. Therefore if water's health, energy, and life force are to be maintained at the highest possible level, then certain precautions must be taken, for this anomalous condition is not only crucial to water's diverse functions, but also to Schauberger's theories and their implementation with regard to the temperature gradient to be elaborated later on.

Dielectric Value

Another important factor is water's dielectric value. The base dielectric value for calculating all other values is based on the permittivity of a vacuum and has a value of 1. Permittivity is the extent to which a substance can be penetrated or traversed by an electric current or charge. The dielectric value of pure water (distilled water) is 81 ($= 9^2$) and is therefore 81 times more effective as a charge separator than a vacuum and almost the highest dielectric value there is.

Specific Heat and Thermal Conductivity

A further life-giving property of water is its high specific heat and thermal conductivity, namely the ability and the rate at which it absorbs and releases heat. According to H. L. Penman's paper "The Water Cycle," water has the greatest specific heat known among liquids ($= 1$) and also has the greatest thermal conductivity of all liquids. "Its great specific heat means that, for a given rate of energy input, the temperature of a given mass of water will rise more slowly than the temperature of any other material. Conversely, as energy is released its temperature will drop more slowly."[6]

This means that a large input or extraction of heat energy is necessary to bring about a change in density and temperature. The lowest point of the curve of the specific heat values for water, however, lies at +37.5°C or 99.5°F (figure 3.7). How strange then, and how remarkable, that the lowest specific heat of this "inorganic" substance—water—lies but 0.5°C (0.9°F) above the normal +37°C (98.6°F) blood temperature of the most highly evolved of Nature's creatures—human beings—a temperature at which the greatest amount of heat or cold is required to change the water's temperature. This property of water to resist rapid thermal change enables us, with blood composed of up to 90 percent water, to survive in a relatively large range and fluctuation of temperatures and still maintain our own internal bodily temperature. If the blood in our bodies had a lower specific heat, they would heat up more rapidly and start to decompose, or quickly freeze if exposed to extreme cold.

When our body temperature is +37°C (98.6°F) we do not have a "temperature" as such. We are healthy and, recalling Schauberger's view, are in an "indifferent" or "temperature-less" state. Just as good water is the preserver of our proper bodily temperature, our anomaly point of greatest health and energy, so too does it preserve this planet as a habitat for our continuing existence. Water in all its forms and qualities is thus the mediator of all life and deserving of the highest focus of our esteem.

Life is movement and is epitomized by water, which is in a constant state of motion and transformation, both externally and internally. Flowing as water, sap, and blood, this life molecule is the creator of the myriad life-forms on this planet. How then could it ever be construed as lifeless in accordance with the chemist's clinical view of water, defined as the inorganic substance H_2O?

This cryptic appellation is a gross misrepresentation. Were water merely the sterile, distilled H_2O as presently described by science, it would be poisonous to all living things. H_2O or "juvenile water" is sterile, distilled water and devoid of any so-called impurities. It has no developed character and qualities. As a young, immature, growing entity, it grasps like a baby at everything within reach. It absorbs the characteristics and properties of whatever it comes into contact with or has attracted to itself in order to grow to maturity. This "everything"— the "impurities"—takes the form of trace elements, minerals, salts, and even smells! Were we to drink pure H_2O constantly, it would quickly leach out all our store of minerals and trace elements, debilitating and ultimately killing us. Like a growing child, juvenile water takes and does not give. Only when mature—when suitably enriched with raw materials—is it in a position to give, to dispense itself freely and willingly, thus enabling the rest of life to develop.

Types of Water

But what is this marvelous, colorless, tasteless, and odorless substance that quenches our thirst like no other fluid? Did we but truly understand the essential nature of water—a living liquid—we would not treat it so churlishly, but would care for it as if our lives depended on it, which they clearly do.

Apart from the actual treatment of water, which I'll discuss later, certain types of water are more suitable for drinking than others, the following being a general classification to be read in conjunction with the table below.

TABLE 3.1. QUALITIES OF DRINKING WATER

WATER TYPE	DESCRIPTION	DRINKING QUALITY
Distilled water	Purest water—contains no other elements	*
Meteoric (rainwater)	Contains some atmospheric gases—no minerals	**
Juvenile (immature water)	Contains few minerals or trace elements	**
Surface water (dams, reservoirs, rivers)	Contains some minerals and salts accumulated by contact with the soil	***
Groundwater	Contains a greater quantity of minerals, salts, etc.	****
True spring water	High in dissolved carbons, carbonic acid, and minerals	*****

WATER TYPE	DESCRIPTION	DRINKING QUALITY
Artesian water	Deep-lying water that may be fresh or saline and contains a variety of dissolved elements, suspensions, and gases	Variable

Distilled Water

This is what is considered physically and chemically to be the purest form of water. Having no characteristics other than total purity, it has a pre-programmed will to unite with or acquire, to extract or attract to itself all the substances it needs to become mature water, and therefore absorbs and grasps at everything within reach. Such water is really quite dangerous if drunk continuously long term. When distilled water is drunk it acts as a purgative, stripping the body of trace minerals and elements.

Meteoric Water—Rainwater

Meteoric water or rainwater, the purest naturally available water, noxious atmospheric pollutants aside, is also unsuitable for drinking long term. It is marginally better than distilled water and slightly richer in minerals, due to the absorption of atmospheric gases and dust particles. As a living organism it is still in adolescence and needs to undergo certain ripening processes in order to be able to be absorbed by the body and be beneficial to it.

Juvenile Water

Juvenile water is immature water, but it is water coming from the ground. It has not matured properly on its passage through the ground. It emerges, perhaps in the form of geysers, from quite a long way down. It has not yet resolved itself into a mature structure and is therefore still of poor quality. It contains a few minerals, some trace elements, and only small quantities of dissolved carbons, but again as drinking water it is not very high grade:

Surface Water

Surface water, from dams, reservoirs, and the like, contains some minerals and salts accumulated by contact with the soil and also from the atmosphere, but generally speaking it is not a very good quality water, partly because it has already been exposed to heavy oxygenation by being in contact with the atmosphere and has also been heated frequently by exposure to the sun, which removes a great deal of water's character and energy.

Groundwater

Groundwater is already much better, often expressing itself as a seepage spring. It is water emanating from lower levels that seeps to the surface after passage along the top of an impervious stratum. It has a larger quota of dissolved carbons, which are the most important ingredient in high-quality water, apart from other trace salts.

True Springwater

True springwater—we shall explore the differences between a seepage spring and a true spring later on—is very high in dissolved carbons and minerals and of the highest possible quality. Its high state of health and vitality is affirmed by its shimmering, vibrant bluish color, which is not evident in inferior waters. Such water is ideal for drinking, if it can be obtained. Unfortunately there are now very few true, high-quality springs left, due to the destruction of the environment.

Artesian Waters

Apart from the above waters, there are artesian waters obtained from bores for wells, which are of unpredictable quality. At times they may be saline and at others brackish or fresh. One can never be sure that well water will necessarily be of drinking quality. Well water probably lies between groundwater and seepage spring water in terms of quality, but most probably can be likened to and classified as groundwater. Once again it depends on how deep the well is and what stratum of water is tapped.

But what are we actually given to drink? This subject of vital interest to us all, which so intimately affects our life, health, and well-being, will be discussed later. Now we must turn our attention to the temperature gradient, which, after the anomaly point of +4°C, is the next most important factor in understanding water and its proper, naturalesque handling.

OTHER FACTORS RELATED TO THE HEALTH OF WATER

The Temperature Gradient

Apart from other factors (some cannot be defined quantitatively), encompassing such aspects as turbidity (opaqueness), impurity, and *quality*, the most crucial factor affecting the health and energy of water is *temperature*, the various aspects of which will be addressed in greater detail later. But first of all a general overview is in order.

Conceived in the cool, dark cradle of the virgin forest, water ripens and matures as it slowly mounts from the depths. On its upward way it gathers to itself trace elements and minerals. Only when it is ripe, and not before, will it emerge from the womb of the Earth as a spring. As a true spring, in contrast to a seepage spring, it has a water temperature of about +4°C (39.2°F). Here in the cool, diffused light of the forest it begins its long, life-giving cycle as a sparkling, lively, translucent stream, bubbling, gurgling, whirling, and gyrating as it wends its way valleyward. In its natural, self-cooling, spiraling, convoluting motion, water is able to maintain its vital inner energies, health, and purity. In this way it acts as the conveyor of all the necessary minerals, trace elements, and other subtle energies to the surrounding environment.

Naturally flowing water seeks to flow in darkness or in the diffused light of forests and shaded streams, thus avoiding the damaging direct light of the sun. Under these conditions, even when cascading down in torrents, a stream will only rarely overflow its banks. Due to its correct natural motion, the faster it flows, the greater its carrying capacity and scouring ability and the more it deepens its bed. This is due to the formation of in-winding, longitudinal, and both clockwise and anti-clockwise alternating spiral vortices down the central axis of the current, which

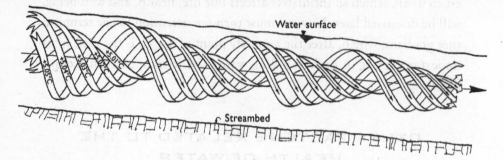

Figure 3.8. The longitudinal vortex showing laminar flow about the central axis. The coldest water filaments are always closest to the central axis of flow. Thermal stratification occurs even with minimal differences in water temperatures. The central core water is subjected to the least turbulence and accelerates ahead, drawing the rest of the water-body in its wake.

constantly cool and re-cool the water, maintaining it at a healthy temperature and leading to a faster, more laminar, spiral flow.

In the illustration above, the coldest water-filaments are always closest to the central axis of flow. Thermal stratification occurs even with minimal differences in water temperature. The central core water is subjected to the least turbulence and accelerates ahead, drawing the rest of the water-body in its wake.

To protect itself from the harmful effects of excess heat, water shields itself from the sun with overhanging vegetation, for with increasing heat and light it begins to lose its vitality and health and its capacity to enliven and animate the environment through which it passes. Ultimately becoming a broad river, the water becomes more turbid, the content of suspended small-grain sediment and silt increasing as it warms up, and its flow becoming slower and more sluggish.

However, even this turbidity plays an important role, because it protects the deeper water strata from the heating effect of the sun. Being in a denser state, the colder bottom strata retain the power to shift sediment of larger grain size (pebbles, gravel, etc.) from the center of the watercourse. In this way the danger of flooding is reduced to a minimum. The spiral, vortical motion mentioned earlier, which eventually led Schauberger to the formation of his theories concerning *implosion,* creates the conditions where the germination of harmful bacteria is inhibited and the water remains disease-free.

In the form of the temperature gradient, the omission of temperature in all hydraulic calculation has resulted in the most devastating floods and the ruination of almost all waterways. While flow velocity, shear force (sweeping force), sediment load, turbidity, and viscosity, to name a few factors, are taken into account in numerous formulae, the temperature gradient, which significantly affects the function of all these different aspects, has so far been totally disregarded in the fields of river engineering, water supply, water resources management, and the condition of water generally.

Apart from variations in its content of organic matter, minerals, and salts—the so-called impurities—water has always been deemed a lifeless, inorganic substance. Therefore, except for certain defined water temperatures required for specific purposes such as cooling and heating, the temperature or variations in temperature of any given water or body of water have hitherto been considered totally immaterial to the behavior of the water itself, since the measured range of these variations has generally been rated too small to be capable of producing any noteworthy effect.

Schauberger defines the temperature gradient, of which there are two forms, as follows:

A positive temperature gradient exists when one of the following conditions is present:
- The temperature of the water decreases and its density increases toward the anomaly point of +4°C.
- The density and temperature increase from freezing and below toward +4°C.
- Ground or water temperatures are cooler than air temperatures.
- Ground temperatures are cooler than water temperatures.

A negative temperature gradient exists when:
- The movement of temperature is away from +4°C, either upward or downward, both of which signify a decrease in density and energy.

In figure 3.7 the directions of movement of these two tempera-ture conditions are shown as two curves delineating the variations of volume and density with temperature. Here it can be seen how with cooling the volume decreases and the density increases, and vice versa with heating. A movement of temperature toward the anomaly point of +4°C always involves a positive temperature gradient, whereas a movement in the opposite direction is indicative of a negative tem-perature gradient.

Both forms of temperature gradient are active simultaneously in nature, but for there to be evolution instead of devolution, the positive temperature gradient must predominate. On both upward and down-ward paths life emerges at the intersection of these two "temperaments," as it were, each of which has different characteristics, properties, poten-tials, and opposite directions of movement or propagation.

Whatever manifests itself as a result of the interaction of these mutually opposing essences depends on the relative proportions between them, which also determines their point of intersection. For example, if the positive temperature gradient is very powerful, then the effect of the reciprocally weaker negative temperature gradient is beneficial and pro-motes the outbirth into physical form of the highest quality substances.

Conversely, if the roles and ratios are reversed and the negative tem-perature gradient is very dominant, then what unfolds as material sub-stance is of inferior worth. For evolution and growth to proceed with increasing quality, vitality, and health, which form is uppermost and at what level of reciprocity their interaction takes place is of absolutely crucial importance, for this not only affects the movement of water, the movement of sap in plants, and the flow of blood in our veins, but also the configuration, structure, and quality of the channels, ducts, and ves-sels surrounding and guiding them.

As it flows, water acts completely differently according to which-ever temperature gradient is in force. In its concentrative, cooling, energizing function, the +4°C-approaching, positive temperature gra-dient has a formative effect. It is a process allowing living systems to be built up, since in water it draws the ionized substances together into intimate and productive contact, for here the contained oxygen becomes passive and is easily bound by the cool carbones,* thereby contributing beneficially to healthy growth and development. The +4°C-deviating, negative temperature gradient, on the other hand, has a disintegrative, debilitative function, for with increasing warming the structure of a given body becomes more loosely knit with a commensurate loss in cohering energy. In this case, due to the rising temperatures, the oxygen becomes increasingly aggressive and reverses its role as co-creator and benefactor, turning into a destroyer and fosterer of diseases and pathogens.

In all waters, forests, and other living organisms the temperature gradient is active in both positive and negative forms. In the natural processes of synthesis and decomposition each has its special role to play in nature's great production, but each must enter on the stage of life at its appointed time. The positive temperature gradient, however, must play the principal role if evolution is to unfold creatively.

Temperature Gradients during Flow

Ever present, the temperature gradient is decisively active in the movement of water and the configuration of flow. Under natural conditions, when water flows down a gradient its flow is affected by a naturally occurring sequence of positive and negative temperature gradients, because in the course of flow the water rhythmically heats up and cools down. How much it heats up, however, is dependent on the degree of friction with the riverbed, the external temperature, and the extent to which the water is directly exposed to the sun. It only requires a very minute change in temperature for water to pick up, transport, or deposit its sediment, and it is the type and duration of the temperature gradient in force that determines which happens and for how long. A negative temperature gradient causes the deposition of sediment, whereas a positive temperature gradient ensures its removal. This pulsation or alternation can be likened to breathing, with a positive temperature gradient representing the inward

---

*Carbones are principally those basic elements and raw materials of carbonous nature, although the term also includes all the elements of the chemist and physicist with the exclusion of oxygen and hydrogen. They are what Schauberger called "mother-substances," as they form the matrix from which all life is created.

breath, the absorbing, material-collecting movement, and the negative temperature gradient representing the outward breath, where the energetically transformed matter is exhaled from the system and deposited.

Apart from the general function of temperature gradients described above, in order to explain the various aspects of temperature-related flow as clearly as possible, each one will be dealt with individually, although by and large in any river or stream all of them are interactive in diverse combinations. Here it is important to understand that every particle of water is directly connected to a particular velocity relative to its specific weight and temperature, a phenomenon described in great detail by Schauberger in his 1930–1931 treatise "Temperature and the Movement of Water."[7]

To give some idea of what is here involved, a series of superimposed water strata with their respective temperatures is shown schematically in figure 3.9a, the coldest stratum flowing over the streambed. Here the velocity curve shows the different distances traveled by the respective water strata in the same period of time, as denoted by the length of the arrows. Relative to the upper stratum, the lowest stratum can be seen to flow far more rapidly due to its greater density and coolness.

At the interface between these various strata, even though the temperature differences may be minimal, there is nevertheless a difference in their relative, temperature-related velocities, the lower stratum sliding forward slightly faster than its immediate upper neighbor. This slip creates a sort of vacuity at the "end," as it were, of the higher-lying stratum, into which the lower stratum rises. In the process vortices are formed at right angles to the current, which rotate on a horizontal plane from the bottom upward, as shown in figure 3.9b. These mix the water but at the same time cool it, because the water temperatures within the center of these vortices are identifiably cooler than those without, the uppermost vortex train manifesting itself as the familiar backward-breaking ripples seen on rivers at the surface. This type of vortex also distributes the lighter-weight sediment and the nutrient material carried by the river from the center toward the sides, as shown in figure 3.9c.

The movement of water can also be further categorized into laminar and turbulent flows, the simplest form of laminar flow being the one shown in figure 3.9a. Turbulence, however, can take the form of longitudinal or transverse vortices. As far as the latter are concerned there are two principal types: the first operates horizontally at right angles to the direction of flow as shown in figures 3.9b and 3.9c; the second, potentially the more harmful, also acts at right angles to the current,

Temperature and the Movement of Water

Every particle of water is connected to a particular velocity relative to its specific weight and specific temperture. When the critical velocity relative to temperature is exceeded, turbulence occurs, which is the NATURAL and AUTOMATIC BRAKE in flowing water.

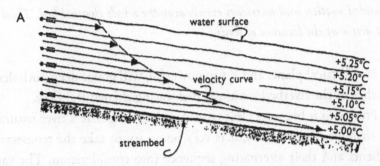

Figure 3.9a. Laminar flow occurs when the internal variations in the water temperature are at a minimum, thus reducing the differences between the respective flow-velocities; there is therefore no turbulence.

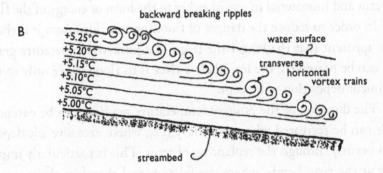

Figure 3.9b. Turbulence in the form of horizontal vortex trains occurs when temperature variations are more marked and the critical velocity of each underlying water stratum is exceeded. In the process of vortex formation, the water is cooled, the flow becomes more laminar once more, and the flow accelerates.

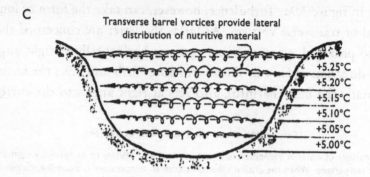

Figure 3.9c. The horizontally acting, paired barrel-vortices transport sediment, suspended matter, and nutrients evenly over the whole channel bed. These are most active at the location of fords.

but on a vertical plane, and if too powerful, will gouge deep potholes or trenches in the riverbed, seriously dislocating the natural flow.

From this it becomes clear that in order to regulate a river naturally, and therefore satisfactorily, it is very necessary to take the temperature gradients and their alternating sequence into consideration. The variations in the temperature of the water body as a whole and in its various parts are so subtle, lying perhaps within a range of 0.1°C to 2.0°C, that contemporary hydraulic engineering has so far never paid the slightest heed to them, the temperature of the water generally being deemed unimportant and immaterial in regard either to the form or energy of the flow.

In order to reduce the danger of flooding to a minimum it is therefore apparent that the longer the reign of a positive temperature gradient can be preserved, the less likely a river is to flood, since only minor sediment deposition will occur.

The duration of the positive temperature gradient can be extended or it can be recreated where necessary (i.e., where excessive silt deposition occurs) through the replanting of trees. This is particularly important at the river bends, where the friction and therefore the warming tendencies are greatest. Species of timber should be planted along the banks, which have a high evaporation rate. In the process of evaporation the sap in the tree is cooled, and because the roots develop underneath the riverbed this cooling effect is also extended to the riverbed and thus to the water as well. Trees therefore act like a refrigerator and help maintain the positive temperature gradient for a longer period.

The key factors here in terms of land and water resources management are, first, never remove forests from the banks of a river; indeed a belt of trees of at least five hundred to one thousand meters wide should be maintained along all riverbanks for the health of the river. Second, rivers flowing through cleared, barren countryside should be densely reforested on both banks in order to cool the water. This will greatly assist in reestablishing healthy flow conditions, restoring the nutrient supply, and recharging the groundwater table through the reestablishment of a positive temperature gradient between the water and the ground in its vicinity.

The Formation of Vortices

Longitudinal vortices, as the name suggests, are aligned parallel to the flow axis of the channel. While these may constitute turbulence according to the meaning of the word, longitudinal vortices have an extremely beneficial function and represent the structuring of those energies required to dislodge and transport sediment, without which all channels will eventually silt up. At the same time they are those vessels that create and enhance the immaterial energies, soul, or psyche of a waterway.

During flow healthy rivers have a natural sequence of clockwise-rotating (right-hand bends) and counterclockwise-rotating (left-hand bends) longitudinal vortices (figure 3.10). In this way the suspended and dissolved carbons, which generally congregate along the banks and bed, are lifted toward the dissolved oxygen, which, in all healthy streams, normally resides in the central flow axis. These fructigenic* carbons react to centripetence. In other words they become very active if moved centripetally, and in this condition are able to bind the fertilizing oxygen, which becomes passive with the cooling centripetence of the central vortical flow, but highly active with warming centrifugence. If the right proportion between centripetence and centrifugence

---

Fructigenic is a term of Schauberger's describing the higher subtle energies responsible for increasing the fecundity or capacity for fructification and fertilization of and by living things.

in the vital longitudinal vortices exists, the most productive interaction between the two opposing substances is also assured. Here they interact not only to increase the energies in the water, but also to augment its carbonic acid content, which is one of the principal constituents of good water. Moreover they create conditions conducive to the propagation of bacteria and microorganisms beneficial to the environment through which the water passes. As the reflection of a primary energy path, the serpentine, meandering pattern of bends in a river is a manifestation of the physical secondary effect. Apart from large, immovable obstacles such as mountains and cliff faces, for example, the course of a river or stream always follows the path in which the energies in a given situation like to move. In some instances it is difficult to say whether the topographical features of a landscape produced the form of the river or whether the river gave rise to the landscape through which it flows (e.g., the Grand Canyon), so intimately connected are the two. Rivers are therefore the mirrors of an unseen flow of energy.

Although there are not many left, a naturally flowing river, undisturbed by modern river engineering, only rarely if ever overflows its banks. In their cool, faster flow down the flow axis, longitudinal vortices clear the channel bed of sediment as well as deepen it, varying this capacity to suit the volume of the discharge. These vortices are also thermally stratified in a laminar fashion. As an example, in figure 3.10 the central core water of such a vortex has a temperature of +5.01°C, very dense and cold, and it moves faster than the more outlying water strata, which become progressively less dense as they warm toward the outside.

According to the Archimedean principle of the denser carrying the lighter, here the densest core water carries the specifically lighter water, because in this inwinding, centripetal, vortical movement the densest water has to flow down the very center.

Apart from cooling the river water, the other principal function of both transverse and longitudinal vortices in naturally flowing rivers and streams, through the generation of backward-breaking ripples (figure 3.9b), is to apply the automatic brake to the descending water. Without this naturally applied brake, the heavy masses of water would overaccelerate, rupture the riverbanks, and cause immense havoc. It is this aspect

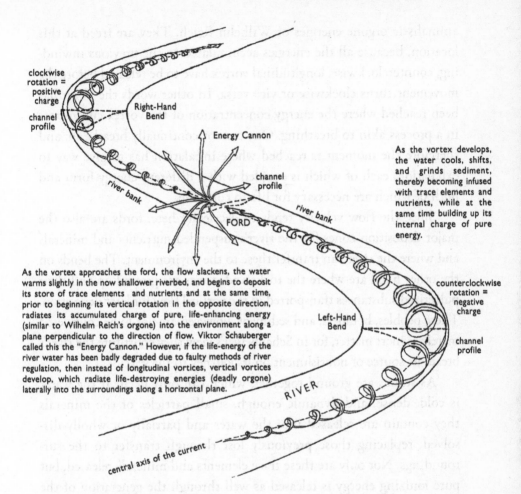

clockwise
rotation =
positive
charge

channel
profile

Right-Hand
Bend

Energy Cannon

As the vortex develops,
the water cools, shifts,
and grinds sediment,
thereby becoming infused
with trace elements and
nutrients, while at the
same time building up its
internal charge of pure
energy.

channel
profile

river bank

FORD

river bank

As the vortex approaches the ford, the flow slackens, the water
warms slightly in the now shallower riverbed, and begins to deposit
its store of trace elements and nutrients and at the same time,
prior to beginning its vertical rotation in the opposite direction,
radiates its accumulated charge of pure, life-enhancing energy
(similar to Wilhelm Reich's orgone) into the environment along a
plane perpendicular to the direction of flow. Viktor Schauberger
called this the "Energy Cannon." However, if the life-energy of the
river water has been badly degraded due to faulty methods of river
regulation, then instead of longitudinal vortices, vertical vortices
develop, which radiate life-destroying energies (deadly orgone)
laterally into the surroundings along a horizontal plane.

Left-Hand
Bend

counterclockwise
rotation =
negative
charge

channel
profile

RIVER

central axis of the current

*Figure 3.10. In the natural flow of rivers, the sequence of longitudinal vortices
will alternate between clockwise and counterclockwise rotations, creating a
riverbed with alternating right-hand and left-hand turns, respectively.*

that forms the nub of Schauberger's initial treatise, "Turbulence,"[8]
deposited under seal by Professor Exner at the Austrian Academy of
Science in 1930.

The location of this current crossover is where the river is shallowest
and where it can most easily be forded. This point is the focus or target
of what Schauberger called the "energy-cannon" (figure 3.10). It is where
the upbuilding immaterial energies of the river are released into the
environment, which, as a form of energy, are akin to the life-endowing,

animalistic orgone energies of Wilhelm Reich. They are freed at this location, because all the energies accumulated in the previous inwinding, counterclockwise, longitudinal vortex have to be released before the movement turns clockwise or vice versa. In other words the point has been reached where the energy concentration of the vortex culminates in a process akin to breathing. One cannot continually breathe in, and therefore the moment is reached where inhalation has to give way to exhalation, each of which is coupled with a different energy form and both of which are necessary for life to continue.

Since the flow velocity tends to decelerate here, fords are also the major deposition zones for the river's suspended nutrients and minerals and where the river can transfer these to the environment. The bends on the other hand are where the rocks and stones are ground up and their pulverized substances transported in the vortical flow for later deposition. These pebbles, boulders, and sediment, however, are not to be considered merely as inert matter, for in Schauberger's view they constitute the river's bread, its source of nourishment on its journey to the sea.

As stones are ground together, which can only occur if the water is cold, dense, and dynamic enough, small particles of the minerals they contain are released into the water and partially or wholly dissolved, replacing those previously lost through transfer to the surroundings. Not only are these trace elements and minerals released, but pure ionizing energy is released as well through the generation of the triboluminescence.*

It is the energetic effects associated with triboluminescence that endow the flowing water with new and immaterial energies, enhancing its character and quality. This is yet another of the many ways in which a river constantly generates and regenerates its energies and vitality, while at the same time imparting them and the necessary nutrients to the environment. If the temperature gradient at the ford is positively

---

*Triboluminescence is an internal glow or luminescence produced when two or more crystalline rocks of similar composition are rubbed hard together or struck against one another, and it is attributed to the energy given off by the electrons contained in the rocks as they return from a pressure-induced, excited state to their rest orbits. As a phenomenon it can occur both in air and under water.

related to the ground temperatures, then these vital nutrients will be absorbed into the ground and the groundwater table thereby recharged and enriched.

A further point of interest in this regard is the origin of the fabled "gold of the Nibelungs," the "Rhinegold" that supposedly lay on the bottom of the Rhine in days of yore and that gleamed during the hours of darkness. This legend is also to be ascribed to the phenomenon of triboluminescence. About 200 to 250 years ago no doubt the water of the Rhine was pure, clear, and translucent enough for people to observe what appeared to be the flashing of gold on the riverbed. Today, however, along with many other rivers, the Rhine is a thick, turbid, grey-green, muddy brew, its life force having been extinguished by modern mechanistic methods of river engineering, so the "gold" can no longer be seen.

DRINKING WATER SUPPLY

The Consequences of Chlorination

Water is *the* issue most crucial to all life on Earth. Water is the lifeblood of our planet, the life-giving fluid in all organisms, plants, animals, and human beings alike, flowing as sap, lymph, or blood. Our very existence is therefore intimately connected with the quality of water available to us. It is vital for our own lives and those of our children that we become seriously concerned not only for the health, vitality, and quality of the water we drink, but also for the source from whence it originates and the treatment it receives, for apart from our own consumption of it, this same water is also used to grow everything we eat. If we want to live in health and happiness, then the living entity—water—should be highly revered and the greatest sensitive care should be given to it.

Today the drinking water supplied to almost all inhabitants of so-called civilized countries is chlorinated and sometimes even fluoridated. The purpose of this treatment is to sterilize the water, to free it of all noxious microorganisms and pathogenic bacteria. Present methods of water treatment and reticulation kill water, however, and bad water or wrongly treated water debilitates, degrades, degenerates, and ultimately destroys those organisms constantly forced to drink it. Science, however, completely overlooks the fact that water—as life carrier—is itself *alive* and needs to be kept in this condition if it is to fulfill its naturally ordained function, for as Schauberger has stated, "Science views the blood-building and character-influencing UR-ORGANISM—'WATER' merely as a chemical compound and provides millions of people with a liquid prepared from this point of view, which is everything but healthy water."[9]

But what does modern, denaturized civilization care, as long as it receives a suitably hygienized, clear liquid to shower with and use to wash its dishes, clothes, and cars? Once down the drain in company with all manner of toxic chemicals and detergents, all is comfortingly out of sight and out of mind. As proof of the efficacy of current disinfective practices, and in justification of their continuance, officialdom usually points out that such water-borne diseases as cholera and typhoid are virtually unknown in all countries where the water is chlorinated.

Thus reassured, the broad mass of the population blithely continues to bask in the luxury of apparently disease-free water in complete ignorance of the perilous repercussions arising from its constant consumption, for what is never stated in official explanations is the cumulative effect this treatment of water has on the organisms forced to drink it. What people do not know is that although the chlorination of drinking and household water supplies ostensibly disinfects them and removes the threat of water-borne diseases, it does so to the detriment of the consumer.

In its function of water sterilizer or disinfectant, chlorine eradicates all types of bacteria, beneficial and harmful alike, so that what arrives at the tap or faucet, while indeed free of every possible organism, is water that has been sterilized to death—in other words, a water corpse. More importantly and more alarmingly, however, it also disinfects the blood (up to 90 percent water) and in doing so kills off or seriously weakens many of the immunity-enhancing microorganisms resident in the body of those organisms forced to consume it.

This eventually impairs their immune systems to such a degree that they are no longer able to eject viruses, germs, and cancer cells, to which the respective host bodies ultimately fall victim. We therefore actually sterilize our blood when we drink chlorinated water, readying ourselves for the onset of disease. Of late there has been an alarming increase not only in hitherto unknown diseases, but in all forms of sickness, cancer in particular.

In view of the fact that our body's water content amounts to 45 liters and that our daily consumption of water in one form or another is about 2.4 liters, just consider table 3.2.

TABLE 3.2. THE WATER CONTENT OF THE HUMAN BODY[10]

The blood plasma	(main blood component)	about 92% water
The human fetus	(our growing physical vehicle)	about 90% water
The blood	(life fluid and nutrient conveyor)	up to 90% water
The human brain cells	(intellect, creativity, behavior)	85% water
The kidneys	(fluid processors and purifiers)	82% water
The muscles	(prime movers of the body)	average 75% water
The body	(our abode on Earth)	71% water
The liver	(metabolism regulator)	69% water
The bones	(structural support system)	22% water
The body's cell fluids	(basis of growth and development)	mainly water

Just imagine what effect the constant drinking of dead or diseased water has on all of these! What happens to the life force essential for healthy growth?

And what are the effects of chlorination? Chlorine is not added to drinking water in vast quantities. On average, it is administered at about 10 parts per million, provided always that the dispensing and metering equipment is properly maintained and monitored. Malfunction, however, can never be ruled out, with the result that over-chlorination may occur more frequently than we are led to believe.

In the process chlorine replaces hydrogen, one of the key elements of the water molecule and present in all carbohydrates and fats, both of which are essential to metabolism in all organic life. One effect of this hydrogen replacement may well be the removal of the hydrogen atoms in the fatty substances surrounding and enclosing the cells, the cell walls, which act as a dielectric membrane and conserve and separate the bioelectric charges responsible for the cells' correct function.

On the other hand, it (chlorine) may also create certain quantities of hydrochloric acid in the blood itself, which as a digestive juice normally resides safely confined within the walls of the stomach, and as a result may add to the overall acidity of the blood, thereby reducing the blood pH* to levels below the normal, healthy level of 7. As a powerful oxidant it also accelerates the metabolic processes of oxidation, on the one hand creating additional heat and on the other consuming oxygen destined for other purposes, and if these occur above the naturally prescribed levels, in most organisms it leads to premature aging.

What more needs to be said, apart from the fact that all these abnormal oxidizing processes cause the dislocation of the natural energy flows in the body, which in turn raises its general temperature, thus placing it in a disease-prone condition—disease after all being the way Nature removes all organisms that are no longer healthy or viable in her scheme of things and that stand in the path of evolutionary progress. In confirmation of chlorine's disease-causing function, a recent study found that in water purification it "produces by-products that cause 18 percent of rectal cancers and 9 percent of bladder cancers."[11]

But this is not where it all ends. Ultimately all these malpractices not only have the direst consequences for the body, but also for its more immaterial attributes, and here we shall quote Schauberger once more: "A particular inner temperature produces a certain physical form which in turn generates the special kind of immaterial energy we encounter in a more or less highly developed form as character. Hence the old say-

---

*pH is the measure of the hydrogen-ion concentration in a given substance and indicates the degree of acidity or alkalinity. Pure water and human blood both have a neutral pH of 7. Above a pH value of 7 alkalinity increases; below it acidity increases.

ing 'Mens sana in corpore sano' (a healthy mind in a healthy body). If the composition of the basic substances of the body should in any way be altered, then the metabolic basis for the further growth of the body must not only change, but its spiritual and intellectual growth and further development as well."[12]

Schauberger saw the proper physical formation of the brain as being crucial to what it produces in the way of concepts, ideas, and behavior, ethical and otherwise. The lower the quality of the physical structure, the more inferior the morals and ethics. In the same way that narrowly spaced annual rings of trees produced high-quality, resonant timber, he saw the production of good thoughts in harmony with nature and, in consequence, good character traits as possible only with a well and healthily grown and developed brain with close-knit windings.

Unwholesome food, poor water, and the resultant slight overheating, in his view, gave rise to the formation of coarse convolutions in the brain's overall structure, creating a brain that was incapable of either functioning intuitively or of comprehending the subtleties of nature's processes. It degenerated into an organ able only to think logically, but never "biologically," never with a living logic aware of natural energetic interrelations and interdependencies. Such a brain could be likened to a poorly designed musical instrument constructed of inferior materials and thus unable to create truly harmonious sounds affecting the world harmoniously. There is plenty of evidence in support of this, for daily we are made aware of the rise in mental afflictions, depression, dyslexia, irrational and brutal behavior, and hyperactivity, to name a few conditions that are affecting more and more people at an increasingly younger age.

It is high time a thorough investigation and highly publicized public inquiry into present methods of water purification be undertaken immediately by an independent body of competent, unsubornable individuals. These should be selected from all branches of science and medicine, including so-called alternative practitioners, whose expertise in some areas far exceeds those of orthodox disciplines. Should its publicized findings recommend the immediate cessation of current practices in water purification, then neither the government nor the respective authorities should be able to continue to hoodwink the population, and they should be forced by the ballot box to take action and undertake the necessary and urgent remedial measures to give us healthy, life-giving water.

The Storage of Water

Whether our water is commercially processed or we obtain it from natural sources, we must care for the very limited stocks of water still available. This means we must treat it in the way demonstrated to us by nature. First and foremost, water should be protected from sunlight and kept in the dark, far removed from all sources of heat, light, and atmospheric influences. Ideally it should be placed in opaque, porous containers, which on the one hand cut out all direct light and heat, and on the other allow the water to breathe, which in common with all other living things, it must do in order to stay alive and healthy.

The present system of bottling water in clear, transparent bottles, for example, detracts from the water's quality because it is exposed to light and heat. When a glass of good water is left out in the sun, little bubbles form on the glass as the carbonic acid, the principal ingredient of good water, is converted into carbon dioxide through increased temperature and light. Like wine, water needs to be kept in the dark in an opaque bottle sealed with a breathing cork, and it is not without reason that good wine is matured in wooden casks.

In terms of what we can achieve personally, we should at all times ensure that our storage vessels, bottles, and tanks are thoroughly insulated so the contained water is maintained at the coolest temperature possible under prevailing conditions. The materials most suited to this are natural stone, timber (wooden barrels), and terra-cotta. Perhaps more than any other material, terra-cotta has been used for this purpose for millennia. Terra-cotta exhibits a porosity particularly well suited to purposes of water storage. This is because it enables a very small percentage of the contained water to evaporate via the vessel walls.

Evaporation is always associated with cooling (vaporization, releasing heat), and according to Walter Schauberger, if the porosity of the container is correct, for every 600th part of the contents evaporated, the contents will be cooled by 1°C (1.8°F). Therefore, if such a vessel is positioned where there is a reasonable movement of air, the water will cool and approach its anomaly point, its state of highest health and "indifference" at a temperature of +4°C (39.2°F).

Another important factor is the actual shape of the container itself. Most storage containers in use today take the form of cubes, rectangular volumes, or cylinders. While these are the shapes most easily and economically produced by today's technology, they have certain drawbacks in terms of impeding natural water circulation and promoting water suffocation.

Due to their rectangular shapes and/or right-angled corners, stagnant zones are created that provide a suitable environment for the propagation of pathogenic bacteria. And since the materials used are generally galvanized iron, fiberglass, concrete, or steel—all impervious materials—the contained water is unable to breathe adequately and suffocates as a result. In this debilitated state it becomes a water cadaver, quickly becomes diseased, and requires further disinfection.

Taking Schauberger's maxim "Comprehend and Copy Nature" as our guide, we should therefore make use of the shapes Nature herself selects to contain, guard, and maintain life, in other words, eggs and their derivations. Should we now make a study of those shapes nature chooses to propagate and maintain life, it soon becomes apparent that the cubes and cylinders have no place in nature's scheme of things. For the safe storage of her vital fluids and materials nature chose eggs and elongated egg shapes such grains and seeds, because she in her wisdom had determined that these produce the optimal results.

Compared with cubes and cylinders, as shown in figure 3.11, these shapes have no stagnant zones, no right-angled corners that inhibit flowing movement. By placing egg-shaped terra-cotta vessels in shaded areas, exposed to air movement, the evaporative cooling effect will be significantly enhanced. Since all natural movement of liquids and gases is triggered by differences in temperature, so too inside the egg-shaped storage vessel, cyclical, spiral, vitalizing movement of the water will be induced. Movement is an expression of energy, and energy is synonymous with life. The external evaporation causes cooling of the outer walls and the water in their immediate vicinity. Being cooler and therefore denser, water becomes specifically heavier and sinks down along the walls toward the bottom, at the same time forcing the water there to rise up

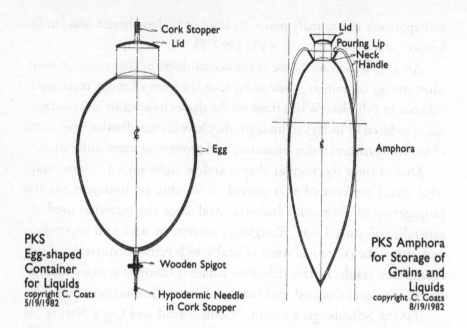

PKS
Egg-shaped
Container
for Liquids
copyright C. Coats
5/19/1982

Cork Stopper
Lid

Egg

Wooden Spigot

Hypodermic Needle
in Cork Stopper

Lid
Pouring Lip
Neck
Handle

Amphora

PKS Amphora
for Storage of
Grains and
Liquids
copyright C. Coats
8/19/1982

Figure 3.11. To hold all-important water, Nature chooses egg-shaped or more elongated seed-shaped containers, which do not have corners that form stagnant zones in which pathogenic bacteria can propagate.

the center and move toward the outside walls. Continual repetition of this process results in the constant circulation and cooling of the contents. As Lord Thurlow said, "Nature is always wise in every part."

CONCLUSIONS

In light of these facts the task we presently face is monumental and far transcends short-term economics, politics, and national self-interest. As a first priority, a program of re-education about water and its essential nature should be inaugurated globally, because the future of water is synonymous with our own. This task requires a committed cooperation from all concerned if there is to be any chance of real success. Here there is no room for competitive behavior or strategies for privatization and profit, for water is a substance nature has made freely available to all and it is our natural birthright. It is imperative that it be made clearly and irrefutably obvious to government that all present economic policies, driven by finance and policies based on economic expediency for short-term gain, will become purely academic and therefore without value or relevance unless the future of the two cornerstones of life on this planet is first assured. Without clean, healthy water and an abundance of thriving vegetation, no economy of any kind is sustainable. These are, and have ever been, the fundamental bases of existence and evolutionary development. Without them all unnatural economies are doomed and so too the countries and peoples that espouse them. Life as we know it will be extinguished. It's as simple as that!

A long-term view is now the imperative of the hour, which means taking appropriate steps now to ensure success and stability in the mid- and long-term future. For this, as Schauberger sought to teach us, a new and far profounder knowledge of nature is necessary, so that whatever is implemented by way of remedial measures will be in harmonious accordance with nature's laws. As never before, we now stand at a watershed and must all act together in consort, for we are, each one of us, inextricably interconnected with the whole, as so eloquently depicted by the great German poet Johann Wolfgang von Goethe in the following poem:

> All things into one are woven, each in each doth act and
>   dwell,
> As cosmic forces, rising, falling, charging up this golden bell,
> With heaven-scented undulations, piercing Earth from
>   power Sublime.
> Harmonious all and all resounding, fill they universe and
>   time!
> Amidst life's tides in raging motion, I ebb and flood—waft
>   to and fro!
> Birth and grave, eternal ocean, ever-moving, transient
>   flow.
> Such changing, vibrant animation, the very stuff of life is
>   mine,
> Thus at the loom of time I sit and weave this living cloth
>   divine.

NEW ENERGY AND
THE IMPORTANCE OF MAVERICK
SCIENTISTS

We are entering a scientific age, but it will be a science which passes out of the impasse which it has now reached and which—having penetrated as it has into the realm of the intangible—will begin to work far more subjectively than heretofore. It will recognize the existence of senses which are super-sensory and which are extensions of the five physical senses, and this will be forced upon science because of the multitude of reliable people who will possess them and who can work and live in the worlds of the tangible and the intangible simultaneously. The mass of reputable testimony will prove incontrovertible. The moment that the subjective world of causes is proven to exist (and this will come through the indisputable evidence of man's extended senses) science will enter into a new area; its focus of attention will change; the possibilities of discovery will be immense and materialism (as that word is now understood) will vanish. Even the word "materialism" will become obsolete and men in the future will be amused at the limited vision of our modern world and wonder why we thought and felt as we did.

DJWHAL KHUL

How often in the past one hundred years has publicly funded institutional scientific research resulted in major scientific and technical innovations? And how often have such discoveries come from the derided and ostracized loner in his "skunk works"? Even a superficial review shows a massive trend in favor of the latter.

Bell and the telephone; Parsons, Tesla, and the turbine; Edison and the electric light and recorded sound; Marconi,

Tesla, and radio; the Wright Brothers and flight; Carl Benz and the automobile; the Lumière brothers and cinema; Otto Mergenthaler and the Linotype machine; Armon Strowger and the automatic telephone exchange; George Eastman and celluloid photographic film; Fritz Haber, the historian who taught himself chemistry and fixed atmospheric nitrogen; Wegener and continental drift; Pollen and automatic fire control; Baird and television; Whittle and the jet engine; Chester Carlson and Xerography; Eckert and Mauchly and the commercial computer; Edwin Land and Polaroid photography; Christopher Cockerell and the hovercraft. Even where the innovator belongs to a recognized institution he or she is often a loner who achieves success by swimming against the currents of orthodoxy, like Alan Turning and the first British computers, or even Watson and Crick, who had been told to drop their study of DNA but continued it as "bootleg" research.

Of course, one can also compile a long and distinguished list of discoveries in institutional science, especially from the great universities like Oxford and Cambridge and especially in important basic fields such as atomic physics and astronomy. But, somehow, it is difficult to draw up a list that carries quite the same diversity, the same romantic air of excitement and innovation and one that has so obviously influenced every single aspect of twentieth-century life so fundamentally. Anyone who switches on the electric light, turns on the television, makes a phone call, watches a film, plays a record, takes a photography, uses a personal computer, drives a car or travels by airplane has the lone eccentric to thank, not institutional science.

RICHARD MILTON,
ALTERNATIVE SCIENCE

4

ROYAL RAYMOND RIFE

THE FATE OF COMPASSION
AND THE CANCER CURE THAT WORKED

Gerry Vassilatos

NOTE: This essay is an excerpt from the book *Lost Science* by Gerry Vassilatos and is reprinted by permission of Borderland Science Research Foundation (www.borderlands.com).

> *The ages cannot kill a truth, and the first man who phrased it will find his echo right down through the centuries.*
>
> PAUL BRUNTON

In the 1930s and 1940s, Royal Raymond Rife revolutionized everything that has been done before or since in high-resolution optical microscopy. . . . He produced direct, economical, electromagnetic cures of cancer, leukemia, and other such debilitating diseases. His work presages a future mankind could have had, where most debilitating diseases were quickly and economically corrected, and where no poisonous drugs, violent nuclear irradiation, and harsh chemotherapeutic "burning" of the patient would be necessary. For such epochal work, he was ostracized, essentially imprisoned in a medical

treatment facility, broken, condemned, and rejected by his
peers. His findings, though printed in reputable publications
and journals, were discredited and ridiculed.

THOMAS E. BEARDEN

LIGHT

There is a constant war being waged that most prefer to ignore. Living
out our days in the joyous sunshine, we rarely choose to glimpse full-
faced into the horrid visage of disease as physicians so often do. Perhaps
it is pain, perhaps fear. Despite our willful ignorance, hideous armadas
of pathogens march through all nations unhindered. These insidious
enemies wage their continual war against the human condition, with a
cruel and merciless deliberation.

Pride and wealth cannot keep these legions away. They are deadly,
having no conscience or allegiance. They are the universal enemy of
humankind, a relentless foe. It is a wonder that nations have not surrendered their petty personal feuds long enough to recognize the common specter. Joining our best forces to defeat this dread army long ago
would have secured major victories for all of humanity.

It has therefore fallen to the sensitive and impassioned few, armed
with vision and swords of light. The independent medical crusaders enter
the battle alone. Their names are seldom seen in major journals any longer.

Figure 4.1. Royal
Raymond Rife with
one of his microscopes

Their private research forever dangles on gossamer threads of grants and endless bureaucratic labyrinths. Yet these are the ones, the men and women who make the discoveries from which cures are woven. Real cures.

They often live on shamefully minuscule budgets, preferring to pour their personal funds into their work. They are the seekers. They are always close on the brink of a possible new development. The important thing is that they are prepared, and they wait for the gracious and providential revelations on which humanity depends. Theirs is the excitement of the chase. Their quest is for "the breakthrough." They are the ones who fill little lab rooms, closet spaces that line university hallways. Their intuitive vision has guided them into research alleys that are too small for the big concerns of profiteering medical agendas.

If these researchers are fortunate, they find an impassioned patron. Perhaps the patron is a sensitive one, whose life has been touched by the sting of tragedy. Perhaps a loved one was lost. Perhaps also in the heat of that pain, the recognition came that gold must be transmuted by passion and devotion before it can cure. These quiet ones who go about their work daily; they have devoted hearts and are driven on behalf of all who bear sorrow over what has been lost.

There was one such man. His discovery gave eyes to the blind. He perfected a means by which humanity's enemies could be detected. His microscope could optically sight viruses, and sight them in their active state. And he developed a means by which a virus, any virus, could be eradicated with the flick of a switch. His medical developments won him no reward or recompense because his research did not fit the desired agenda.

THE UNIVERSAL MICROSCOPE

There were predecessors to the prismatic marvel—the super microscope—of Dr. R. Raymond Rife, but no equals. Others had designed and used oil immersion lenses, dark-field illuminations, and deep ultraviolet light, each holding part of the secret for optically magnifying infinitesimal objects. But the design that Rife developed outpaced all of these.

And yet it is doubtful whether you have ever heard Rife's name. Reasons for mass-forgetfulness run deep. *Truths have been kept from you.* Only a careful and relentless study of the past will relinquish secrets purposefully and cunningly buried. The information is safely nestled in dust-laden libraries that few nowadays venture to peer into. Perhaps you will recognize why his name has been blotted out of the historical records before we reach the end of this amazing biography.

Rife began as a research pathologist. A medical crusader of the very highest qualifications, his was a heart filled with but one goal: the eradication of disease. Rife recognized first and foremost that successful medicine relies on vision, on light. What we cannot see we cannot battle. An unseen foe is impossible to destroy. Therefore his first quest was to secure a vastly improved system of microscopy. Once he could see, once *all* could see, then the pursuit of medical knowledge could again move forward. An armada of equipped seers could assail the foe on every shore.

Rife's study of microscopy detailed every component and premise that tradition had presented to the twentieth century. The creation of a super microscope would run counter to every physical law and restriction of the previous two centuries.

Rife wanted to develop super microscopes capable of seeing viruses. His aim was to chart and catalog them, understanding that they represented a deadly foe that exceeded bacilli in their destructive assault on humanity. His quest now began. He reduced the fundamental premises by which microscope design had developed, analyzing each separate component and premise.

FOCUS

Optical designers had been adding ever more complex components to the design that began with Antonie van Leeuwenhoek. Lenses were compounded to lenses, crowns were added to compounds, crowns were added to crowns—the complexity was frightful. Simplifying the problem necessarily led Rife back to the study of optical geometry and the comprehension of simple ray divergence.

Rife thought on these ancient principles. An ideal magnifying system is a geometric construction of extreme simplicity. Diverging light rays can magnify any object to any magnification. Given a strongly divergent light source and a great enough distance, one can theoretically magnify the indivisible! This is the principle that underlies projection microscopy. Rife realized that the projection microscope represented the best and simplest means of magnifying infinitesimal objects. One simply needed to discover a means by which a vanishingly small, brilliant radiant point could project divergent rays to the surface of any material speck. No virus, however indistinct and cunning, could hide from such an optical magnifier.

The theoretical design of a microscope relies purely on geometric principles. Actual materialization of these principles requires material manipulations, since geometric rays and light rays are significantly distinct. What is a microscope essentially? What is achieved in a microscope with light rays? The notion is quite simple. Take divergent rays from a vanishingly small point of brilliant emanations and allow them to pass through any spectrum, which is to be viewed. Light from this encounter is then made to diverge as far apart as possible in a given space. Geometrically it is possible to divert rays from a vanishingly small point out to an infinite distance.

This geometric construction produces unlimited (ideal) magnifications. Provisions toward this ideal goal require that the point radiant source be tiny and brilliant enough, the specimen be close enough to the radiant point, and the image diverging space be very long. The geometric divergence of the point light source is the magnification factor. But geometry is an idealized reality. And ideal geometry encounters significant frustrations when implementing light in inertial space.

The most basic type of microscope is the projection microscope. It is the simplest system employed to greatly magnify the most infinitesimal objects. In the more common version, light is made to pass through a tiny specimen. Light from the specimen is forced to diverge across a long space by means of a very small focusing lens. Rays from this lens cross, diverging and expanding across a long space. This widely divergent beam is then projected onto frosted glass. The viewing of images derived through these means is indirect, but provides superior magnifications with ultrahigh resolutions.

Previously, laboratories required compact units capable of close personal manipulations. The development of fine optical microscopes became frightfully complex when more powerful but compact models were required. The idea behind the compact microscope is the physical compression of a lengthy projection to the limited space of a compact tube, delivering the shrunken design to customers who wish to conserve space. The "problem" with compact optical microscopes lies in bending the necessary wide beam through a small space. The "trick" in a compound microscope is to keep the image rays from diverging prematurely between lens stages.

The long expansion space required for divergent beam magnification had to be "folded" and "convoluted" within imaging tubes. Large numbers of lenses performed this duty. Being thus convoluted by lenses to achieve magnification, images produced by most expensive bench microscopes were inherently limited. Since the diverging image in these microscopes is "interrupted" within a greatly shortened space by means of several optical stages, the microscopes cannot produce great magnifications with either clarity or brilliance.

Each optical stage continually bends the image until a tremendous effective divergence is achieved. The effects are dramatic, but the necessary stages introduce optical resistances by which magnifications are inherently limited. Fundamental problems with white light alone complicated the problems that designers faced. Breaking into spectral components, each color refused to focus in exactly the same point. As a result, chromatic aberrations blurred every image.

The light-crossing action of each lens brought widely diverging light beams into the ocular lens. It was pivotal that these rays be parallel. Images lost most of their radiant power against the tube walls before arriving in the final ocular. Therefore more corrective lenses were added in the beam path to bend the light back from the tube walls. Differences when light traveled between lenses and air introduced more aberrations. Batteries of corrective lenses, crowns, and components so loaded the light path with crystal that images lost their original brightness. These horrendous optical problems were never completely solved, despite the high cost of these instruments.

All of these optical horrors were the result of an old tradition that compelled designers to maintain familiar outward forms. The projection microscope is so simple and potent, one wonders why newer designs had not been developed with as much dedication and zeal. It was the outward form that compelled the convolution of projection microscope simplicity, detracting from the excellence of magnified images. What was really lacking in optical microscopy was the development of true, tiny radiant points of monochromatic light. These diverging ray sources could produce novel and economical projection microscopes.

The numerous optical components of most excellent laboratory microscopes are configured to prevent image splitting, image incoherence, and other optical aberrations. When light and glass are introduced the geometric ideal suddenly becomes severely limited. The optical ideals fall short of the geometric ideal.

Geometric rays do not fade with infinite distance. Light rays do. Geometric rays do not blur at their edges with increasing divergence. Light rays do. Geometrically magnified lines do not diminish in their intensity. Light images do. A successful optical approximation to the geometric ideal would produce a superior microscope. Rife decided to manipulate all the possible variables in order to approach, as closely as possible, each part of the ideal geometric construction. If such a feat could be accomplished, he would successfully bridge the gap between optical and electron microscopy.

POINTS

To be sure, numerous individuals had accidentally discovered enormous magnification effects while experimenting in completely different fields of study. A magnifying system that magnified much smaller infinitesimals than viruses appeared in 1891. Nikola Tesla developed a remarkable carborundum-point vacuum lamp and made an accidental observation that opened a new world of vision to science.

Tesla began inventing single-wire vacuum lamps for purposes of illumination. These were large glass globes powered by very rapidly impulsed currents. The impulse currents made the single supported wires glow to white brilliance, melting them. Because they were impractical for public use, he sought to alleviate this condition by using special crystals, for which high melting points were required. An assortment of such materials was poised at the single-wire termination. When electrified, they suddenly became radiant.

His experiments included the use of diamonds, rubies, zircons, zirconia, carbon, and carborundum. He found it possible to blast the natural gems after a few seconds of electrification. Before exploding, each of these crystalline terminations released puzzling patterns of light across the globe surface. This symmetrical pattern of points attracted Tesla's attention. They appeared when the current was turned on for just an instant.

Moreover, Tesla noticed that the brilliant fixed points of light remained in fixed positions each time he applied the current. Equally astounding was the fact that each material portrayed distinctive point symmetries on the glass enclosure. The most resilient and successful crystalline material was carborundum, which he ultimately adopted for practical use. This too gave its characteristic point symmetry across the globe.

Tesla was not sure what he had discovered. He intuitively surmised that these point patterns of light somehow revealed the crystalline structure of the excited materials. He also used the geometrical construction to obtain his deductions. His thoughts turned to the internal crystal conditions. As electrically charged particles were propelled and ejected from the carborundum, they formed infinitesimal points that diverged and impacted the inside of the globe housing the carborundum. These brilliant points of light were always of the same symmetry because the ejected particles were passing through a fixed grating, a crystalline grating.

He theorized that this fixed pattern represented the greatly magnified crystalline symmetry. This simple apparatus was the world's first point-electron microscope. The phenomenon responsible for the defined projection of crystalline spaces is referred to as field emission. The remarkable X-ray photography of Max von Laue already permitted the sighting of crystalline atoms. In this scheme a thin crystal point was placed at a critical distance from an X-ray source. Entering and passing through the crystal slice, divergent X rays produced a greatly magnified image of crystal atoms on photographic negatives.

The result of von Laue's experiment was astonishing, but was a purely geometric consequence. Divergent rays from a vanishingly small radiant point can theoretically magnify equally small specks to immense size. But while both Tesla and von Laue produced wonderful results with particle-like emissions, the practical achievement of these ideals was diminished when using optical light rays.

Emile Demoyens (in 1911) claimed to have seen extremely tiny mobile specks under a powerful optical microscope, but only at noon during the months of May, June, and July! Colleagues thought him quite mad, but Dr. Gaston Naessens has comprehended why these specific time periods permitted such extreme viewing. During these seasonal times the noonday sunlight contains great amounts of deep ultraviolet light. The shortened wavelengths provide a sudden optical boost, permitting the observation of specks that are normally invisible.

Progress in optical science seemed limitless and free. It was anticipated that no limit could bar humanity from viewing the very smallest constituents of matter. But when the physicist Ernst Abbe challenged the high hopes of optical science by imposing certain theoretical limits on optical resolution, all these hopes seemed to dissolve. Abbe claimed that optical resolution depended entirely on incident light wavelengths, the limit being one-third of the light wavelength used to illumine the specimen. According to Abbe, the extreme ultraviolet light of 0.4 microns wavelength could not be used to resolve the details of objects smaller than 0.15 microns.

This theoretical "death-knell" discouraged most optical designers of the time. He claimed that the resolution of optical microscopes was restricted from 1,600 to 25 diameters and, because of this, developing newer optical microscopes was a futile pursuit. Since resolution is the ability of a magnifying instrument to identify details and ultrafine levels of internal structure, the Abbe limit imposed a serious halt on the development of newer optical microscopes.

Continual medical progress rides entirely on the excellence of its instruments. In the absence of new and excellent optical instruments of greater precision, medical progress grinds to a screaming halt. When this happens, in the absence of "true" vision, academics write endless papers. True knowledge, reliant on vision and experimentation, is replaced by unfounded speculation.

Others conceived of electron microscope designs, taking advantage of the Abbe restriction for lucrative purposes. These developers were not good planners, failing to recognize that electron microscopy would place equally grave limitations on biological researchers. Electron beams kill living matter, and by magnifying images only after killing them, no living thing could ever be observed in natural stages of activity through electron microscopy. But, if money were to be made, then "all was possible." Despite the protests of qualified medical personnel, RCA continued its development with Vladimir Zworykn at their helm.

Electron microscopy, rationally impelled by the Abbe limit, became the new goal of young financiers. Despite the protests of major researchers, RCA continued its propaganda campaigns. This technological imposition, were it developed into a marketable product line, would severely handicap the work of every medical researcher. Pathologists would be literally forced to accept the limitations of the anticipated electron microscope.

Bracing themselves for the announcement of mass-produced electron microscopes, corporate researchers prepared themselves for the laboratory adaptations they would be forced to adopt. Manuals were already being distributed. They would be unable to watch progressive activities in the boasted "highest magnifications ever achieved." Before RCA reached the goal, however, others had already challenged the capabilities of electron microscopy. The unexpected development temporarily threw RCA off balance. The competitors had challenged the Abbe limit and seemed to be optically working their way into realms in which RCA had claimed "exclusive" rights.

DEEP VIOLET

Vibrating above the deep ultraviolet range were the X rays of von Laue's projection microscope. But this realm was not good for pathologists since X rays would only reveal the structure of crystalline substances. Some designers went ahead and built soft X-ray microscopes. These devices placed heavy requirements on the preparation of specimens. X rays passed right through specimens and killed them if they were alive to begin with. The very best X-ray images of tiny specimens required organism-killing metallic stains. Biologists needed to see their specimens in the living state.

While engineers at RCA were scrambling to take the competitive edge and seize the new market, several designers of ultramicroscopes began to successfully challenge the Abbe limit. Ultramicroscopes constructed by Louis C. Graton and Ernest B. Dane, Jr. (of Harvard University) succeeded in developing resolutions of 6,000 diameters with magnifications of 50,000 diameters.

Dr. Francis Lucas of Bell Telephone Labs developed a modified version of this system in which a maximum magnification of 60,000 diameters was developed. Not only did this work significantly reduce the theoretical limits set by Abbe, but the ultra-microscope that Lucas designed actually empowered Bell Labs to compete with RCA in the microscope field. Rife had previously achieved resolutions of 6,000 diameters with magnifications of 50,000 diameters. And now, Rife believed he had a means by which these preliminary feats could be greatly outperformed. The Abbe limit, a theoretically perfect expression, was dissolving before the new empirical evidence.

Of course, RCA ultimately outdid the propaganda campaign for their own electron microscope system, wiping out the optical systems of both Bell Labs and Harvard University. Nevertheless, independent researchers preferred those ultraviolet microscopes to any system that RCA could market. The ultraviolet microscopes were attractive because they permitted life-active observations. Consequently, pathologists were not impressed by the extra magnifications of electron microscopy.

OBJECTIVE

Ultraviolet light for ultra-microscopes is an absolute necessity. The successful operation of any such device depends on deep UV rays. Monochromatic ultraviolet sources prevented many of the familiar optical aberrations common to optical microscopy. Blurring and fringe degeneration when passing through the optical resistance of lenses would be minimal. The ultraviolet source would also need to be of the shortest possible wavelength in order to approach the geometric ray idea.

All optical components in the ultra-microscope would then have to be composed of pure quartz crystal in order to flawlessly transduce the deep ultraviolet rays. Even the specimen slides were made of thin quartz glass. The ultra-microscopes of Dane, Graton, and Lucas used as few lenses as possible, being virtually pure projection microscopes.

According to Lucas, resolution one-tenth of the illuminating light wavelength was obtained. This broke the so-called Abbe optical restrictions by an order of 300 percent; the resolution was brought up to 0.05 microns. How was this possible? Dane and Graton further stated that far greater resolution could be obtained through lenses than claimed by their manufacturers. The reason for this? As long as the manufacturers had accepted the theoretical limits there was no incentive toward progress in the field. No one bothered to find out!

The ultra-microscopes demonstrated beyond question that lenses do in fact surpass theoretical limits. The manufacturers, eager to maintain credibility in the academies, had simply endorsed whatever the physicists wrote. Equally as significant was the fact that each of these ultra-microscopes did not require the fixing of specimens before viewing. The embodiment of each ultra-microscope gave new drive to researchers who wished to see live pathological stages in tissue cultures. The systems were immediately demanded and obtained by numerous serious research institutions on both sides of the Atlantic.

Certain highly respected researchers came to believe that the most basic laws concerning physical light were fundamentally flawed. Perhaps light was of an entirely different nature than supposed. This, they mentioned, was why the Abbe limit was such a distorted mathematical expression. Light was not what the physicists declared it to be. This is why Abbe's assessment was so obviously flawed. But what other assumed truths were holding back fresh discovery? Empirical observations now replaced the theoretical piles with discoveries that were once termed "unlikely" by qualified authorities.

When researchers realized the great cost that the Abbe limit had so long imposed on microscope designers, they began challenging every known theoretical limit pertaining to their fields of study. Every scientific premise was questioned during the astounding decade of the 1930s. Every applicable optical rule was again subject to fresh questioning, which epitomized the spirit of the renewed scientific mind. New vision filled the researchers, challenging the inertial world again. The most significant effect of these new ultra-microscopes was a renewed questioning process. Now pathologists and biologists alike were given instruments with which to peer into the most infinitesimal natural recesses.

With the ability of medical researchers to look into the deepest pathogenic lairs, new cures for ancient maladies could be effected. The war was on, and fresh crusaders came to the battlefield armed with light. Curiously, the lines of battle brought two distinct groups to fight the same foe. Unfortunately, one group desired all the glory and crushed its more sensitive brothers.

The Rockefeller Institute extended their campaign by highlighting the efficacy of electron microscopy, securing the sale of their new units. The RCA cash flow was unrestricted now. Electron microscopy coupled its forces with the pharmacological industry, producing its line of allopathic medicines. Those who took upon themselves the inquisitorial profession, rather than the profession of truth, found themselves drowning in seas of new developments that their business-minded patrons wished to eradicate. Independent university researchers maintained their poise as the prime recipients of fresh and astonishing discoveries that shook the medical world. This would not long be tolerated by the growing pharmaceutical monopolies and trusts who wanted total domination of the field.

"MAGNIFICANT"

The encroaching economic depression of the time had crushed the general populace. Rife had been designing and assembling ultraviolet precision microscopes of superior quality from 1920 onward. He had planned to build a far superior instrument: the super microscope. The design was based on theoretical considerations developed during his preliminary experimentation in optics. Now this work was abruptly terminated. Finding himself out of employment, Rife sought the ordinary work of those who are in need. Humbled and not proud, he temporarily sought a salary in less intellectual venues.

Hired as private chauffeur to Henry H. Timkin, a wealthy and philanthropic motor magnate, he gradually won both the respect and willing ear of his adventurous employer. He could not keep his wonderful dream to himself. On long journeys to boring boardroom meetings, Timkin engaged Rife in detailed discussions on his medical work. Rife eagerly entered these discussions with an enthralled candor that caught his employer quite by surprise. The seriousness and integrity of the man did not catch Timkin by surprise. Timkin, recognizing quality when he heard it, listened.

Rife's great stature was not hidden, despite his humble position. And when he spoke of these designs and research goals, the very air began to brighten around him! He mentioned regret at having to postpone his work, but was very sure that all would turn out well. What he had shared with Timkin was enormous; it was inspiration of the purest kind. When Timkin and his business partner Harry Bridges realized exactly what Rife hoped to achieve, they made a resolute decision to arrange financial support for the work at hand.

Timkin and Bridges created an endowment fund to finance Rife and his astonishing research. Rife was delighted, delighted to tears. An emotional man, he promised that no one would be disappointed. He would work until he achieved success. A laboratory was constructed on the Timkin estate grounds (Point Loma, California), and Rife set to work with a fury that surprised those who lovingly surrounded him. Timkin and Bridges were taken aback by the rapidity with which Rife completed each design that he began. His efforts were relentless, a true inspiration to the equally loving people who supported his research. It was very apparent that the pensive and gentle doctor was serious in the extreme.

Rife aggressively pursued and achieved what had not been done in the field of ultra-microscopy. His mind turned to the method he had conceived so many years before. The dream that Rife had originally received was now in view, that of looking for more light. He decided to try filling the entire objective with cylindrically cut quartz prisms. There would be no difference in the refractive index from start to finish along the optical path. Quartz prisms would "open out" each ray convergence, maintaining strictly parallel ray cadence. An increased ray content being thus returned to the ocular, the image would be brilliant in appearance and of high resolution.

This configuration of quartz prisms caused the rays to "zigzag" in twenty-two light bends. The internal optical path was entirely composed of twenty-two quartz blocks, fitted snugly to lenses. Now, specimen-emergent light would launch out in parallel paths through quartz prisms, being magnified only when they reached each quartz lens. This optical tracking method would ensure the brilliance of the emergent image.

A second optical innovation was added to this brilliant configuration. Rife decided to use a phenomenon by which strong specimen-entrant light stimulates internal fluorescence in the specimen. Pumping the specimen with brilliant ultraviolet-rich light would shift the divergence point into the very heart of the specimen rather than beneath, forcing the specimen to radiate its own brilliant ultraviolet rays.

Here was a true, vanishingly small radiant source with which to illuminate the specimens: they themselves would become the radiant source! This concept was truly sublime, since the very infinitesimal particles themselves were now made to radiate brilliant and diverted rays. This scheme was truly original from the very start. Rife then designed a system by which selected portions of the ultraviolet spectrum could be split and directed into the specimen using a polarizer. Turning this component of the system would allow each specimen to brightly fluoresce in its own absorption spectrum, the infinitesimal specks radiating their own maximum brilliance.

Theoretically, it was possible to magnify these brilliant specular rays to any degree. But a secondary monochromatic ultraviolet ray would perform an unheard of wonder. When combined with the brilliant internal fluorescence of the specimen, this secondary ultraviolet addition would heterodyne the light. This meant that light pitches from the specimen would be raised far above their original values. At such short

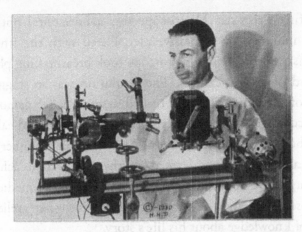

Figure 4.2. Privately funded during the Great Depression, Rife went to work on designs for what he called a "super microscope," which used cylindrically cut quartz prisms along the ocular path to maintain a strictly parallel ray cadence, which resulted in greater brilliance and resolution in images.

wavelengths, the resolving power of this device would be incredible.

An additional monochromatic deep ultraviolet beam was mixed with the florescent radiance of spectrums, producing an astonishing visual sharpness of otherwise invisible objects. The illumination scheme and the tube-filled cluster of quartz prisms (designed to maintain the spectrum-emergent rays in absolute parallels) were now brought together. Rife claimed that these parallel lines were within one wavelength of accuracy, an astonishing claim.

He soon created a small ultra-microscope whose fundamental mode of operation violated the supposed laws of optics. This design outperformed all previous ultra-microscopes. So astonishing was this feat that the Franklin Institute, in rare form, published a long and detailed series of articles concerning the developments of Rife. They were also given several of these units for preservation, where they remain to this day.

This microscope was different, totally different, because it could see viruses in their active stages (not only in their dormancy) with magnified clarity. Rife's Prismatic Microscope surpassed the theoretical limits that were possible for optical microscopy in 1930, giving unheard of resolutions of 17,000 diameters; three times the resolution developed by Lucas.

The first Prismatic Microscope was a horizontal optical bench assembly, mounted on a massive pier. Fitted with the finest photographic instruments, Rife's microscope took breathtaking photographs at unheard of magnifications. The resolution was so staggering that members of research institutions rushed to watch Rife's demonstrations.

His accomplishments were extolled by the entire medical establishment on both sides of the Atlantic. An incredible number of professional research publications devoted lengthy articles to his achievements. His findings were duplicated and reported by leading medical institutions whose names are well known, throwing into sharp relief our general lack of knowledge about his life's story.

Rife, humble enough to have worked as a chauffeur, had been raised from obscurity to fame, from shadows to light. The man's genius was only equaled by his upstanding character. The Timken family adored him. The Rife laboratory was completely equipped with the very finest apparatus money could buy. Rife methodically undertook new research ventures. Incredible new biological discoveries followed him in every direction. Now, with this "ultravision," he was able to peer with his colleagues into dimensions unheard of. These discoveries quite often challenged accepted biological and medical notions.

Rife wanted to create an institute in which he could train younger specialists in the operation of these wonderful ultra-microscopes. Mass production of the devices would be ensured. They would become fixtures in every professional laboratory. Money was not the aim of this research. Monies were already secured. Rife had a singular goal and demonstrated the passion associated with his quest.

He developed seven different models of this initial projection-type Prismatic Microscope in quick succession. The horizontal projection format was converted to a more compact vertical orientation, serving the needs of pathologists and biologists and practical laboratory settings. Several of these wonderful prismatic models may be seen in the various archival films and photographs taken in Rife's laboratories.

If the Rife Prismatic Microscopes outperformed every standard laboratory microscope in being able to discern and photograph virus particles in their active state, the Universe Microscope surpassed all

*Figure 4.3. Royal Rife's
Universal Microscope*

previous markers. In 1933 the creation of the Universal Microscope accorded resolutions in an astounding excess of 31,000 diameters, with magnifications in excess of 60,000 diameters.

Using technically precise photographic enlargement techniques, Rife was able to provide 300,000-diameter magnifications. His calculations indicated that an ultraoptical projection microscope giving clarified magnifications of 250,000 diameters would be possible. After photographic enlargement, there would be no limit to the optical viewing power unleashed for researchers.

Rife had succeeded in breaking the "vision barrier." Those familiar with optics and attainable optical precision states claimed that magnification effects could not be obtained with ordinary principles of light. Beyond simple optical parameters, other light energies become more active in such devices. The focusing process is radionic in effect, using the penetrating Od* luminescence. The stimulation of special retinal modes releases the anomalous perception with its reported ultraoptical

---

*Od refers to the Odic forces studied by Wilhelm Reich. His work was suppressed in the early twentieth century but is making a comeback in several fields, from healing to free-energy technology.

magnifications. Careful examination of the Rife Ultra-microscope reveals tubes filled with quartz prisms, identical in basic use as the patented radionic analyzers of T. G. Hieronymus (Lehr).

Viruses remained absolutely invisible to the eye when cultures were searched with the then-standard Zeiss dark-field (oil-immersion) microscope. Rife's Prismatic Microscopes were immediately obtained by Northwestern University Medical School, the Mayo Foundation, the British Laboratory of Tropical Medicine, and other equally prestigious research groups. These models produced magnification and resolution up to 18,000 diameters.

A space composed of brilliant light, where mind-illuminating light merged with light in the eyes, was now opened before him. Fields, all of light. The new vision would be unstoppable. No cloak of invisibility would protect the foe now. Soon everyone would see, and the armadas of death and shadow would be vanquished. The spoils of this war would flood humanity with indescribable treasure. Light and life would again be unleashed in a world where shadow and death had reigned far too long. The immediate task of cataloging viral pathogens had begun.

QUEST

With the new Prismatic Microscope models, both Rife and Dr. A. I. Kendall (Northwestern University Medical School) were able to observe, demonstrate, and photograph "filterable" pathogens (viruses) in 1931. Moreover, they were perhaps first to discern the transition of these bacilli from dormancy to activity over a specific period of time. Freshly made cultures were sampled at specific stages, revealing fixed periods of quiescence and activity.

An initial tissue substrate was prepared in which bacillus typhosus was cultured. After several days' growth, samples of this lethal culture were filtered through a fine triple-zero Berkefeld "W" filter. This filtration process was repeated ten times. When viewed under the best available laboratory microscopes a turbidity was seen, but no organisms whatsoever appeared.

Under the Rife Prismatic Microscope, polarizer adjusted, the bacilli in this sample fluoresced with a bright turquoise blue coloration. Two forms were observed, taking the researchers by surprise. Long, relatively clear, and nonmotile bacilli were found alongside a great population of free-swimming ovoids, granules of high motility. The motile granules glowed in a self-florescent turquoise light at a magnification of 5,000 diameters.

These motile forms were transferred to a second fresh substrate and allowed to grow for days. The same filtration process was performed. When sampled randomly over a four-day period, the filtered specimen revealed something remarkable. Rife and Kendall observed relative quiescence, with bright turquoise ovoids at one end. The implication was enormous. Exact transition periods were therefore determined with precision and the entire process photographed through special attachments designed by Rife.

At specific intervals of activation, the clear bacilli discharged the turquoise motile forms into the culture. These blue ovoids were the real cause of the disease. The long and clear bacilli were only hosts. Transitions back and forth (between clear host-dormancy and motile turquoise granules) were observed and reported in the professional journals. These findings were corroborated first by Dr. A. Foord, chief pathologist at Pasadena Hospital, and later confirmed and reported by Dr. E. C. Rosenow at the Mayo Foundation (in 1932). The Rife Prismatic Microscope was quickly earning its reputation.

Soon, other specimens were obtained and analyzed by the team. Active poliomyelitis cultures were studied, the virus successfully isolated, identified, and photographed in 1932 by Rife and Kendall. In these cultures streptococcus and motile blue forms resembling typhosus were recognized. These last reports were immediately transmitted to the Mayo Foundation and duplicated by Dr. E. Rosenow. Dr. Karl Meyer (director of the Hooper Foundation for Medical Research, University of California) came to the Rife Research Laboratories with Dr. Milbank Johnson, examining and corroborating the stated results. The impossible and anomalous became fact. Bacilli could act as virus carriers. Furthermore, poliomyelitis victims evidenced a startling degree of typhosus-like associated virus.

Frightening implications came when comparisons between the Prismatic Microscope and the Zeiss scopes were made. All of the previous studies made with Zeiss scopes returned negative results. Such reports flooded the literature. The filtrates had been maintaining their cloak of invisibility for years. Professionals, bereft of this clarified vision, were concocting numerous speculative explanations for the appearance of these disease states. The vacuum produced by lack of visible evidence was producing erroneous theories. Many highly qualified persons, in absence of the foresight dictating that they should know better, steadfastly maintained that victims of certain diseases were suffering from internally developed conditions.

The Rife ultramicroscope was about to trigger a war on viruses. Because of the self-fluorescent "staining" method, Rife observed live specimens exclusively, a distinguishing feature of his technology. The florescent coloration of each pathogen was cataloged, a historic endeavor. Tuberculosis bacilli appeared emerald green, leprosy was ruby red, E. coli was mahogany colored—each one was wickedly deceptive in the pretty colors. The degree of precision demonstrated in Rife's catalogs bears the unmistakable mark of genius.

We can view him at work in the archival movies. Photographic arrays of all kinds may be seen in this footage, including that taken using the professional Scandia 35 mm movie camera with which he made stopaction films of viral incubation periods. Rife made sure to document every discovery. It was novel at the time to document every image on movie film as well as in still shots. He methodically went through every possible pathogenic specimen, photographing the deadly families.

The Prismatic Microscope was piercing into new shadows. Rife recognized unknown virus species everywhere. And then he turned his vision into the deepest shadow. He looked at the dreaded disease: Cancer.

In the absence of fact, in the absence of vision, researchers developed contradictory theories concerning cancer and its development. These contradictory theories were eventually codified in the professional literature, a self-neutralizing amalgam of conjecture. Researchers were forced to examine the biochemical effects, and not the cause, of cancer. Most could not imagine what would drive cells into the bizarre and abnormal cycles common to cancer tissues. There was certainly "no visible cause."

Rife began obtaining a wide variety of malignant tissues in 1931. The full power range of the first Prismatic Microscope was turned on these tissue samples with a vengeance. Rife was a master pathologist, one whose techniques could be observed in his cinematic presentations. Was he seeing correctly? What were these motile forms, glowing with a beautiful violet-red coloration? They moved swiftly through his field of view.

Rife obtained increasingly diverse tumors from wider and more diverse clinical sources. An amazing twenty thousand of these tissue samples were obtained and cultured. Incubating and culturing each of these required care and time. Absolutely sterile conditions were maintained. He employed several groups of large high-pressure steam autoclaves. No question of contamination could exist in this setting. Specimens removed from these cultures were filtered through unused triple-zero Berkefeld porcelain, mixed with triple-distilled water.

Examination of each separate sample under the Prismatic Microscope revealed a consistent truth. There they were again! Always the same violet-red presence. He called it the BX virus, finding it present in every case of cancer in humans. Were these same violet-red motile forms the very cause of cancer? Colleagues were able to verify these findings only when using his microscope. Both Rife and Kendall successfully demonstrated the isolation of the BX virus to more than fifty research pathologists associated with the most noteworthy institutions.

Many writers of medical theory had already postulated that there were some cancer cases that were viral in origin, but they never cited that there are agencies universally causing the cancer; there was speculation, the endless papers, the lectures, the theories—talk and more talk. Rife *saw* the universal calls of cancer. There, in full sight, was the positive proof. In case after unmistakable case, Rife found the very same agency at work, and it was always the same violet-red motile forms. It mattered not where the tissue materials came from. Independent acquisition of tissue samples were obtained by others who then verified these findings in distant laboratories. They were using the Rife Prismatic Microscopes.

Rife succeeded in isolating the BX virus in 1931. He cultured this evil spawn and proceeded to demonstrate its incubation and activation periods. Transferring BX virus from culture to host, and from host to culture, all became routine. One hundred and four separate transfers were successfully made with various BX strains. Rife witnessed the appearance of another related viral cancer-causing strain, the BY virus, found to be a much larger strain of the sarcoma group. Demonstration of the infection in an incubation process was subsequently affirmed by other professionals.

Rife assembled high-speed movie cameras in order to clock the periods of BX virus activity. When the film ran out and was developed, he and all his colleagues could watch the deadly dance. He stepped back for a moment and surveyed the photographic evidence, flickering on the wall. These wicked, damned, wiggling specks!

Rife now watched in horror as the malignant act was revealed before his eyes at high speed. BX virus infection required special "weakened" physiological states. Contracted as a flu-type infection, the virus incubated in its host physiology for a time. When specific detrimental physiochemical states were compounded, the virus stirred into activity.

Stimulating the rapid proliferation of cell division, the BX virus forced the host body to manufacture the nuclear material needed in order to survive. Tumors were found to be sites where BX viral colonies were rampant. Occasionally there were persons who demonstrated spontaneous remissions. These were exceedingly rare cases where antibodies actually drove off the attacking virus. Most persons could not summon this degree of response. Once the virus took control of cellular integrity, death was imminent. There had to be a means for destroying this enemy. There had to be a light.

BREAKTHROUGH

Others, working in distant laboratories, did not claim the same success. Why not? Because, using the over-celebrated electron microscopes, they could not see what Rife saw. The truth regarding the BX virus was that electron microscopy could not image it at all. What had occurred in the other research labs became clear again to the man with eyes to see.

In preparing specimens for an electron microscope, technicians "kill" the tissue specimens. The process involves placing the specimen in a high-vacuum chamber. Bombardment of the specimen with metal ions is the "staining" procedure. The thin metal film then gets highly projected. The electron spray is directed into this prepared specimen and is then magnified by successive intense magnetic field coils. At that point, images are watched on a phosphorescent screen or photographed directly.

Electron microscopy mishandled frail viruses. It mishandled the frail BX virus, destroying it during each preparation process. The same ritual was repeated a hundred times with the same negative results. Few of these technicians could comprehend why this virus did not appear on their viewing screens. Overconfidence in the RCA system blocked common reason. Electron microscopy does not resolve frail viruses because they are shattered and dissolve during the preparation stage. Quite recently the search for the HIV virus evidenced frustration again because of these inherent limitations of electron microscopy.

The BX viruses cavorted and wiggled boldly before Rife's eyes. But how to destroy them? To find an immunological tool for each of these represented an enormous task, a project that would take centuries. Humanity did not have that much time to wait. No, some other, more universal means had to be developed by which this, and all of the other pathogenic forms, could be dissolved.

Protozoa and bacteria of all kinds could be destroyed by exposing them to special ultraviolet spectra. Perhaps the BX virus would succumb to such exposure. Rife had to know, and, to this end, he began a long and arduous search, looking for spectra that could destroy virus cultures.

Rife discovered that deadly viruses actually thrived in the radiations of specific elements. Radium and cobalt-60 were the notable ones. Dormant viruses became virulent in these energetic emanations. The horror filled him again. Medical practice was attempting to cure cancer with these very radiations! There had to be some light spectra that destroyed the viral activity altogether. He searched through the periodic table. Electrified argon and neon also brought intensified virulent activity from dormant cultures. He actually used argon lamps to grow virus-infected tissue cultures with greater rapidity. But there had to be a spectral range that killed these terrible death agents.

No light seemed to have any effect on their crystalline structures. This is why it was possible for him to view viral activity under intense light in the first place!

Then he thought of crystals. How would we destroy a crystal? What do chemicals do to germs—dissolve them, take them apart, shatter them?

He had done this very thing in 1917 with protozoa and large bacteria. He knew it was possible to shatter these kinds of pathogens by the application of a sudden electrical impulse. His early attempts with small radio transmitters and simpler microscopes had proved to be somewhat effective. He used Telefunken output tubes to produce the impulses. Operated by a small generator, this simple device projected fifty radio-frequency watts to his samples.

His original inspiration applied to larger pathogens. It therefore needed no excessive frequency; shortwave was sufficient. It was certainly possible to interpolate the necessarily superhigh resonant pitch needed to shatter any microbe. But viruses? How high would this pitch need to be? If not attainable, could he use some much lower harmonic of this fundamental at greater power levels? Could he find the lethal pitch for every found pathogen?

Equipment was quickly assembled. He needed a generator of extremely short-duration electroimpulses. Direct current electrical "spikes" of quick duration, when applied to a gas-filled discharge tube, would project electric rays toward an infected site. The tube could not simply be a high vacuum. That would release dangerously penetrating X rays. X rays would stimulate the BX strain into increased activity. No, the projection tube required a very light gas, one whose response was almost instantaneous. The gas Rife desired would be one whose mass would in no way interfere with the impulses.

Hydrogen was used in special high-power thyratrons: quick-acting high-voltage switches used in diathermy machines and (later) in radar systems. Old X-ray tubes often failed in their operation because they became filled with hydrogen and helium mixtures. Such X-ray tubes were generally discarded. His new projector was one such old X-ray tube. He tested its output, adjusting the excitation circuit so as not to release even soft X rays. The tube glowed, a good sign. This meant that there was sufficient gas for the release of electrical rays. Rife set the polarity so the tube would pulsate electropositive spikes of a specific duration.

Power was ready. Pathogens cavorted boldly in view. Poised at the Prismatic Microscope, he fired the X-ray tube. Turning to the tuning dial near the specimen, he would know the lethal pitch by watching the pathogens. When these "exploded" he would mark the setting. If this method worked, then he could methodically correlate each lethal pitch with its pathogen. Soon, a catalog of lethal pitches would be amassed. *With this Rife could wage victorious wars against every disease in existence.*

Rife swept through the diathermy range, which he calculated should vibrate these viruses to pieces. Empirical evidence always contradicted the theoretical. Quite below the calculated extreme frequencies, the BX virus suddenly dissolved. He switched off the transmitter and sat there quite amazed. The scene in the microscope was unreal. Not a fraction of a second at the lethal pitch and the specimen was reduced to a globular mass. The viruses were stuck together in shattered fragments! He had successfully "devitalized" them.

Fine-tuned lethal frequencies now filled his catalog. With great precision Rife determined every lethal pitch as planned. Armaments of light against legions of shadow. Analysis of the electropositive impulse showed that its radiance was penetrating, intense, and unidirectional— more like invisible light rays of pure electrical force. What then was this strange lightlike power? Experiments proved that virus cultures were absolutely incapacitated, congealed, and destroyed by the electropositive impulse. The power of an extreme form of light? Had such light ever been seen before?

This energy had been accidentally generated in 1872 by Elihu Thomson and Edwin Houston, but they had generated electrical rays, not waves. Unidirectional electric impulses of great power radiated electric rays, not waves. These rays penetrated all kinds of matter, stone and steel alike. The resultant sparks could be drawn from every insulated metal object in the large building in which the experiment was performed.

Later in that century, Nikola Tesla accidentally observed the same electric ray production. He studied the phenomenon exclusively, developing impulse generators and electric ray projectors. He referred to this phenomenon when speaking of electric rays that evidenced a light-like nature. Rife had rediscovered this phenomenon. Tesla spoke of his own "millimeter rays," mentioning their "bacteriocide" value.

FORTRESS

Whereas the destruction of virus cultures on a quartz slide was easily accomplished, the destruction of pathogen cultures in human hosts was not. Rays had to penetrate through skin, musculature, and bone—which offered considerable resistance. Rays might lose their original accurate pitch in this transit, destroying the intended action altogether.

Fortuitously and strangely, the pathogens were found to be some two thousand times weaker than body cells. This meant that pathogens could be destroyed by the radiant impulse method without harming the patient. How sublime. Pathologists had treated microorganisms as chemical systems for a century, working overtime in order to find each specific chemical-dissolving agent. *Rife's method treated all germs as mechanical systems, dissolving them with vibrations.*

Rife himself had been exposed to the instantaneous blast without harm. When adjusting the rates to annihilate ordinary viral infections, he noticed that he became drowsy and tired for a few hours. Determining the cause of this as the resultant toxin release after infected agents were coagulated, he recognized the need for a detoxifying agent. Physiology had to be prepared for the curative impulse. Exposure would release large amounts of toxic pathogen fragments into the bloodstream all at once. The ray cure had to be metered in doses. Body tissues had to be flooded with special fluid electrolytes to aid the enhanced and rapid elimination of toxins.

To stimulate the deepest potential shattering action, the patient had to be based in a "carrier field": an electrical body permeation in which the impulse light rays would penetrate into and through every body cavity. Superficial exposures would not completely cure the patient. This light ray energy had to penetrate the body completely. Rife conceived a method whereby patients could be enveloped in a harmless body-permeating electric field of acoustic frequency, while the intense electroimpulses of short duration would be simultaneously projected. In this manner, efficacious electroradiant impulses could shatter specific pathogens throughout the infected body with no harm to the patient.

Rife used two banks of oscillators with which to generate his primary and secondary impulse fields. Acoustic generators supplied the primary field of "immersion." A diathermy machine was coupled to a powerful transmitting amplifier to provide the shattering impulse. Two radiant energies were thus employed to destroy pathogens in vivo.

Rife discovered that virus cultures were not safe from the radiant impulses of the special ray tube. Fixed to the lethal pitch of the single pathogen, the rays were unerring in their message. Selectivity was the hallmark of the Rife curative method. Choosing the lethal pitch for one of these, the others would remain unharmed. The target, however, was utterly destroyed.

Rife tested the lethal effective distance of his rays, determining the safe placement of patients from the radiant source. Pathogen cultures did not seem "safe" anywhere near the device. Arranging the tube at one end of his laboratory, Rife moved cultures away from the radiant tube in staggered distances. In a final amazing experiment, he took cultures away from the laboratory in sealed containers. *It was found that radiant tube emanations operated effectively on viral cultures up to an eight-mile distance!* Metal cabinets did not protect viral cultures from the deadly ray effects either, being ray conductive. Even when locked in aluminum cabinets, the entuned light-like rays destroyed their pathogens wherever they were found.

This represented a major medical discovery of greatest value to all humanity. This principle actually made curative broadcast possible. Entire populations could be electrically "vaccinated" from single monitored sites.

The world potential of this system was enormous. Now the outbreak of epidemics could be controlled without the time-consuming need for individual inoculations. The radiant lethal message would eradicate specific pathogens in a simple broadcast. The constant monitoring of socially prolific germ populations could be maintained by continual public health broadcasts.

CONQUEST

He ran his entire staff through varied frequency exposures. Infections of all kinds each dissolved before the ray. Rife was able to isolate the pathogens of infection and destroy them with the mere turn of the dial. The specificity of the Raytube device was so precise that singular germ strains could be individually mass-targeted. People could be cured by the flick of a switch!

Firing the tube in the lab provided a continual source of inoculation. After a time, so little toxicity was present in staff members' bodies that the drowsy effects were never again encountered. They did not contract any illnesses, not even a cold.

After a time, Rife rarely used gloves when handling the viral specimens. Furthermore, neither he nor his technicians ever contracted any of the diseases handled. The Raytube "inoculated" them all against every disease. He reported these findings to the community.

Rife, a research pathologist, never used these devices in medical practice. Other physicians desired the units for their own purposes, recognizing the potential for curing human suffering. Dr. Lee De Forest supervised the design and assembly of many oscillator components for the Rife system. W. D. Coolidge (General Electric) himself willingly sent Rice hundreds of X-ray tubes, which were altered with a mixture of hydrogen and helium by Rife and his technicians. These improved tubes were tested so that they would project only the desired electroimpulse rays.

Hearing of these wonders, numerous physicians began requesting that smaller, more portable units be designed. Soon, Rife Raytube devices were being assembled and given to physicians for limited use in their practices. When properly operated, these devices returned successful reports, effecting complete eradications of infections and cures of various conditions. There were never any adverse reports concerning the Rife Raytube instruments, nor would there be.

The Rife frequency devices were bringing about a therapy revolution. Strep throat* could be cured in an instantaneous exposure, seated in a physician's office. A specially designed gargling solution was given to remove the resulting toxicity from the site.

In 1934 Dr. Milbank Johnson, chief medical director of Pacific Mutual Life Insurance Company, established a therapy center for cancer treatment in Scripts Castle, San Diego. A staff was brought together from specific institutions including Dr. G. Dock (professor of medicine, Tulane University), Dr. C. Fischer (Children's Hospital, New York), Dr. W. Morrison (chief surgeon, Santa Fe Railroad), Dr. R. Lounsberry, Dr. E. Copp, Dr. T. Burger, Dr. J. Heitger, Dr. O. C. Grunner (Archibald Cancer Research Committee, McGill University), and Dr. E. C. Rosenow (Mayo Clinic). Rife functioned as a general consultant in matters of system therapy.

Using a Rife Raytube system, the team received the cancer and tuberculosis patients. Fifteen cancer patients, each pronounced hopeless by medical experts, arrived at the clinic. Each evidenced progressive states of the disease. A few patients were ambulatory. Treatments with the Rife Raytube method were routinely applied. The dream was becoming real.

Recognizing the critical condition of their patients, it was decided that exposure time would be raised to three minutes in duration. It was discovered that exposure could not be repeated without long rest periods. These critically ill patients could not withstand the extreme resultant toxicity released into the system as the BX viruses were shattered. Emotional depression often resulted until the ray dose was safely analyzed. The team conferred hourly to assess the progress of each patient. Excessive exposure to the rays could result in severe lymphatic infections and blood poisoning. Therefore, three-minute treatments were

---

*It was from strep throat, easily curable today, that George Washington died after a life of surviving war, dysentery, influenza, malaria, mumps, pleurisy, pneumonia, rickets, smallpox, staph infections, tuberculosis, and typhoid fever.

repeated every third day, given that rest periods were necessary in order for blood detoxification to occur.

Soon, the ray had done its work on the once-terminal victims. Constant blood and tissue samples revealed no BX viral presence in these now fortunate individuals. In sixty days' treatment time, and after examination by several physicians, each was released as cured.

The patients were under surveillance, and no relapses occurred. The treatment was revolutionary. The results were thrilling and complete. Moreover, they were confirmed by a special medical research committee of the University of Southern California. Three more clinics were opened with Johnson as general medical supervisor. Other participating physicians included Dr. James Couche, Dr. Arthur Yale, Dr. R. Haimer, and Dr. R. Stafford. Clinics were operated between 1934 and 1938 and had such a remarkable number of cures that it is difficult to list them all without simply reprinting the Rife files. Each of these cases was sent out and corroborated by other (nonparticipating) physicians.

In 1939 Rife was formally invited to address the Royal Society of Medicine, which had recently corroborated his findings. He was requested with great enthusiasm to bring all possible films, slides, and apparatus. Dr. R. E. Seidel reported these findings and formally announced the Rife Raytube System therapy for cancer in the *Journal of the Franklin Institute*.[1]

The formation of the Ray Beam Tube Corporation was announced, through which several models would become available to the medical world within a short time. Highly skilled hospital staff members and leading physicians were very receptive to the proliferation of this therapy. *Here was a new means for controlling and eradicating any kind of disease by the flick of a switch.* This therapy would inadvertently challenge pharmacological methods, raising human standards to a new and lofty height. The dream seemed ready to materialize.

INQUISITION

Rife found both himself and his staff members under a strange series of attacks by unknown agencies. During this time, and under very mysterious circumstances, Johnson was found dead in a hospital bed, after being admitted for a relatively minor health problem. In 1939, the local chapter of the Medical Association proceeded to bring Rife to the San Diego Superior Court, but lost their case. Rife could not be charged with malpractice, being a research pathologist and designer of medical instruments.

This repugnant offense unmasked the heinous resentment behind which many powerful individuals had previously been camouflaged. The court action itself caught Rife completely unawares. A visionary, his entire life had been dedicated to humanity. Alleviating human suffering was his life's goal. Here now was strange evidence that factions within the medical establishment were actually mobilizing against proven therapeutic methods. Cancer itself and other equivalent maladies were being cured. Why then the assault?

Growing opposition from deeper factions of the Medical Association brought pressure on staff members of the Rife Treatment Clinic. Threats and other unprofessional pressure tactics forced members to leave the team in quick succession. In campaigns clearly waged to malign Rife and his findings, the Medical Association assailed remaining participants in the clinic until Rife stood alone.

Deeper than the verbal show of malignancy by other colleagues was the horrifying and insidious motivation, the implication behind the attack. Why would anyone wish to destroy such a great achievement? Who was betraying civilization in this crucial instance? Of all betrayals and of all personnel, who in the medical profession would seek the eradication of such monumental discoveries? Rife's mind reeled under the weight of these thoughts. This was no mere resistance to a new idea in a time of ignorance. Pasteur had experienced the same indignity. No. This was willful, calculated resistance taking place in what was an allegedly enlightened time.

Horribly shocked at the entire scenario, Rife literally became unhinged in court. Trembling and weeping, he could not come to terms with the sheer hatred and vehemence exhibited by his antagonists. "Why . . . why are you doing this?" he repeated. The prosecution could not have produced a better effect. Seeing this weakness as the very means by which to eradicate Rife and his discoveries, they continued to attack him openly. Calling him continually to the witness stand, they succeeded in destroying this frail-hearted man of humble greatness. In short, the prosecution forced his total collapse.

Dr. James Couche was compelled to desist operating Rife therapy clinics under threat of malpractice. The American Medical Association ruled that no society member who maintained the use of the Rife Raytube System would be permitted to continue medical practice in the United States. Morris Fishbein, a major AMA stockholder, treasurer, censor, controller, and editor of *The Journal of the American Medical Association,* extended his legal arm to inform each member of the Rife team of the impending legal process. All Raytube units would be recalled, impounded, and destroyed by a federal court order, under penalty of fines and imprisonment.

LIGHT

All participants willingly returned their Rife units except Drs. James Couche and Arthur Yale. These two surgeons later stated that for twenty-two years after this action, they continued to successfully treat and cure thousands of patients with the Rife Raytube devices that they had secretly maintained. Yale published a large and concise chronological account of patients treated and cured in his practice throughout that twenty-two year period. Notwithstanding the fact that 60 percent of severe cancer cases brought to him were medically inoperative, incurable, and hopeless, Yale confirmed that all of these persons were still alive and living happy, full lives.

The Rife microscopes challenged RCA and its lucrative electron microscope. The Rife Raytube system replaced the highly lucrative pharmacological model then in operation. Such developments did not inspire the entrenched corporations who were making enormous amounts of money treating chronically ill patients. Rife developed a therapeutic means that worked. This is all too evident by the rage of those who assailed him. Implicated in the suppression campaign were corporate trusts and government agencies, including the FDA and the National Institute of Health.

Systematic eradications like this one speak of social control on a vast and hideously deep-rooted scale. The notion that disease proliferation was permitted so that the moneyed interests of pharmaceutical companies could be pursued unchecked is too terrible to consider. However, federal officers impounded the entire Rife laboratory all too late. Several faithful technicians had already purloined every piece of the priceless equipment, taking laboratory components and valuable documents across the Mexico border, where they remain to this day. (John Crane, Rife's partner and chief engineer, maintains the priceless surplus there.)

Fishbein, the editor and chief censor of the AMA, saw to it that Rife's name was stricken from all previous publications, that no professional journal would dare publish anything by Rife, and that no mention would ever be made of Rife's achievements in formal proceedings.[2] Fishbein's actions, inescapably linked with the pharmaceutical trusts, were all too conspicuous.

Social control has become a dominant theme since the Second World War. Modifying and regulating social thought through both legal and financial means has brought natural discoveries and true technological developments to a standstill. World-changing discoveries can be made but not proliferated. Cures for diseases can be proven but not implemented.

Has the world now entered a new barbaric and vulgar time where medical wonders have become a regulated property? The historical evidence would seem to say yes. Balancing profit against cost, it is clear that outright cures are far less profitable than exceedingly prolonged and profit-effective "treatments."

Is the honor once laid on the development of wondrous cures to be forever shunned and the cures themselves suppressed willy-nilly by corporate interests? Is compassion for suffering humanity and concern for the elevation of human living standards on a worldwide scale no longer a major medical imperative?

In the past, medical discoveries were never questioned or resisted. They were always looked on as absolutes: if a medical cure for disease was found, it was embraced and implemented with miraculous providence. Not even the most ruthless financier would have dared to interrupt the flow of medical discoveries in past times.

When the records are actually examined, we find a staggering disproportion. How is it that medical research of the nineteenth century, not very well equipped or funded, provided definitive cures that have become part of the established medical landscape today, while contemporary medical research, better equipped and superfunded, has not produced a single cure of equal social importance in the last forty years?

Medical authorities have stated that "no means has been found by which viruses may be destroyed." Recent evaluations of "recaptured" Rife Raytube units contradict this statement. Rife treated germs as mechanical systems, not chemical systems. His essential discovery was that vibrations could kill pathogens by the flick of a switch. As we have stated earlier, a single such device could be easily tuned to destroy all deathly pathogens. *His is the only device that can destroy viruses.*

UCLA Medical Laboratories, Kalbfeld Lab, Palo Alto Detection Laboratory, and San Diego Testing Laboratory had all stated that the Rife Raytube System was absolutely safe to use. The FDA went out of its way to publish and maintain federally directed rulings on the Rife Raytube System, refusing to make further statements concerning its historically proven effectiveness and thousands of cured cancer cases.

A great gathering of esteemed colleagues of the medical and research professions came to honor and support Rife when the entire court affair was over. Friends who had been too frightened to stand and fight at his side were now smiling, with their drinks elevated. But the man who was asked to stand and receive the honors saw through the charades.

The seer saw the thick shadows that enveloped the professionals and other dinner guests. Armadas of pathogens were drumming their war drums again. Soon on the march, they would devastate humanity once more. It seemed that not one of the esteemed guests cared.

Rife developed and implemented a revolutionary cure for cancer and other diseases, something that no contemporary medical research

Figure 4.4. Dr. Royal Raymond Rife at the San Diego Tribute on May 6, 1938

group had been able to do, nor would be able to do, so great had the powers of suppression become in that post–World War II era. The cheers and accolades rang on, and standing ovations for Rife lasted for more than fifteen minutes. The now frail and ghostlike discoverer, however, looked away—far off and away. He was searching through the shadows, searching in his own darkness—for new light.

5

T. TOWNSEND BROWN

THE SUPPRESSION
OF ANTIGRAVITY TECHNOLOGY

Jeane Manning

T. Townsend Brown was jubilant when he returned from France in 1956. The soft-spoken scientist had a solid clue that could lead to fuelless space travel. His saucer-shaped discs flew at speeds of up to several hundred miles per hour, with no moving parts. One thing he was certain of—the phenomenon should be investigated by the best scientific institutions. Surely now the science establishment would admit that he really had something. Although the tall, lean physicist—handsome, in a gangly way—was a humble man, even shy, he confidently took his good news to a top-ranking officer he knew in Washington, D.C.

"The experiments in Paris proved that the anomalous motion of my disc airfoils was not all caused by ion wind." The listener would hear Brown's every word, because he took his time in getting words out. "They conclusively proved that the apparatus works even in high vacuum. Here's the documentation . . ."

Anomalous means "unusual"—a discovery that does not fit into the current box of acknowledged science. In this case, the anomaly revealed a connection between electricity and gravity.

That year *Interavia* magazine reported that Brown's discs reached

Figure 5.1. T. Townsend Brown

speeds of several hundred miles per hour when charged with several hundred thousand volts of electricity. A wire running along the leading edge of each disc charged that side with high positive voltage, and the trailing edge was wired for an opposite charge. The high voltage ionized air around the disc, and a cloud of positive ions formed ahead of the craft and a cloud of negative ions behind.

The apparatus was pulled along by its self-generated gravity field, like a surfer riding a wave. *Fate* magazine writer Gaston Burridge in 1958 also described Brown's metal discs, some up to thirty inches in diameter by that time. Because they needed a wire to supply electric charges, the discs were tethered by a wire to a Maypole-like mast. The double-saucer objects circled the pole with a slight humming sound. "In the dark they glow with an eerie lavender light."

Instead of congratulations on the French test results, at the Pentagon he again ran into closed doors. Even his former classmate from officer candidates school, Admiral Hyman Rickover, discouraged Brown from continuing to explore the dogma-shattering discovery that the force of gravity could be tweaked or even blanked out by the electrical force.

"Townsend, I'm going to do you a favor and tell you, don't take this work any further. Drop it."

Was this advice given to Brown by a highly placed friend who knew that the United States military was already exploring electrogravitics? (Sleuthing by American scientist Dr. Paul LaViolette uncovered a paper trail that led from Brown's early work, toward secret research by the military, and eventually pointed to "black project" aircraft.)[1]

HARASSMENT

Were the repeated break-ins into Brown's laboratory meant to discourage him from pursuing his line of research?

Brown didn't quit, although by that time he and his family had spent nearly $250,000 of their own money on research. He had already put in more than thirty years seeking scientific explanations for the strange phenomena he witnessed in the laboratory. He earlier called it electrogravitics, but later in his life, trying to get acknowledgement from establishment scientists, he stopped using the word *electrogravitics* and instead used the more accepted scientific terminology "stress in dielectrics."

No matter what his day job, the obsessed researcher experimented in his home laboratory in his spare time. Above all he wanted to know, Why is this happening? He was convinced that the coupling of the two forces—electricity and gravity—could be put to practical use.

An arrogant academia ignored his findings. Given the cold-shoulder treatment by the science establishment, Brown spent family savings and even personal food money on laboratory supplies. Perhaps he would not have had the heart to continue his lonely research if he had known in 1956 that nearly thirty more years of hard work were ahead of him. He died in 1985 with the frustration of having his findings still unaccepted.

The last half of his career involved new twists. Instead of electrogravitics, at the end of his life he was demonstrating "gravitoelectrics" and "petrovoltaics"—electricity from rocks. Brown's many patents and findings ranged from an electrostatic motor to unusual high-fidelity speakers and electrostatic cooling, to lighter-than-air materials and advanced dielectrics. His name should be recognized by students of science, but instead it has dropped into obscurity.

Too late to comfort him, some leading-edge scientists of the mid-1990s resurrected Brown's papers, or what they could find of his papers.

EXTRAORDINARY CURIOSITY

Thomas Townsend Brown was born March 18, 1905, to a prominent Zanesville, Ohio, family. The usual childlike "why?" questions came from young Townsend with extraordinary intensity. For example, his question, "Why do the [high voltage] electric wires sing?" led him later in life to an invention. His discovery of electrogravitics, on the other hand, came through an intuition. As a sixteen year old, Brown had a hunch that the then-famous Coolidge X-ray tube might give a clue to spaceflight technology. His tests, to find a force in the rays themselves that would move mass, led to a dead end. But in the meantime the observant experimenter noticed that high voltages applied to the tube itself caused a very slight motion.

Excited, he worked on increasing the effect. Before he graduated from high school, he had an instrument he called a gravitator. "Wow," the teenager may have thought. "Antigravity may be possible!" World-changing technological discoveries start with someone noticing a small effect and then amplifying it.

Unsure of what to do next, the next year he started college at California Institute of Technology. Even then his sensitivity was evident, because he saw the wisdom of going forward cautiously—first gaining respect from his professors instead of prematurely bragging about his discovery of a new electrical principle. He was respected as a promising student and an excellent laboratory worker, but when he did tell his teachers about his discovery they were not interested. He left school and joined the navy.

Next he tried Kenyon College in Ohio. Again, no scientist would take his discovery seriously. It went against what the professors had been taught; therefore it could not be.

He finally found help at Denison University in Gambier, Ohio. Brown met professor of physics and astronomy Paul Alfred Biefeld, Ph.D., who was from Zurich, Switzerland, and had been a classmate of

Albert Einstein. Biefeld encouraged Brown to experiment further, and together they developed the principle that is known in the unorthodox scientific literature as the Biefeld-Brown effect. It concerned the same notion that the teenager had seen on his Coolidge tube—a highly charged electrical condenser moves toward its positive pole and away from its negative pole. Brown's gravitator measured weight losses of up to 1 percent.

(In 1974 researcher Oliver Nichelson pointed out to Brown that before 1918, Professor Francis E. Nipher of St. Louis discovered gravitational propulsion by electrically charging lead balls, so the Brown-Biefeld effect could more properly be called the Nipher effect. However, Brown deserves credit for his sixty years of experimentation and developing further aspects of the principle.)

Brown's 1929 article for the publication *Science and Inventions* was titled, bluntly, "How I Control Gravity." The science establishment still turned its back. By then he had graduated from the university, married, and was working under Biefeld at Swazey Observatory.

His career in the early 1930s also included a post at the Naval Research Lab in Washington, D.C., staff physicist for the navy's International Gravity Expedition to the West Indies, physicist for the Johnson-Smithsonian Deep Sea Expedition, soil engineer for a federal agency, and administrator with the Federal Communications Commission.

As his country's war effort escalated, he became a lieutenant in the Navy Reserve and moved to Maryland as a materials engineer for the Martin Aircraft Company. Brown was then called into the Navy Bureau of Ships. He worked on how to degauss (erase magnetism from) ships to protect them from magnetic-fuse mines, and his magnetic minefield detector saved many sailors' lives.

PHILADELPHIA EXPERIMENT

The "Philadelphia Experiment" that Brown may or may not have joined in 1940 is dramatized in a popular movie as a military experiment in which United States Navy scientists tried to demagnetize a ship so that it would be invisible to radar. According to the account, the ship and its crew dematerialized and rematerialized—they became invisible and later returned from another dimension.

Whatever the Project Invisibility experiment actually was, Brown was probably an insider as the navy's officer in charge of magnetic and acoustic mine-sweeping research and development. However, later in life, Brown was said to be mute on the topic of the alleged Philadelphia Experiment, except for brief disclaimers. He told friend and entrepreneur Josh Reynolds of California, who made arrangements for Brown's experiments in the early 1980s, that the movie and the controversial book *The Philadelphia Experiment* by William L. Moore and Charles Berlitz were greatly inflated.[2] He apparently did not elaborate on that comment.

Reynolds spoke on a panel discussion at a public conference (dedicated to Brown) in Philadelphia in 1994, along with highly credentialed physicist Elizabeth Rauscher, Ph.D. Rauscher theorized that the Philadelphia Experiment legend grew out of the fact that certain magnetic fields can in effect "degauss the brain"—cause temporary memory loss. If the huge electrical coils involved in degaussing a ship were mistuned, the sailors could have felt that they "blinked out of time and back into time."

Going back to 1942: Brown was made commanding officer of the navy's radar school at Norfolk, Virginia. The next year he collapsed from nervous exhaustion and retired from the navy on doctors' recommendations. More than his hard work caused his health to break down; he had suffered years of deeply felt disappointments because his life's work—the gravitator—had not been recognized by scientific institutions that could have investigated it. The final precipitating factor for his collapse was an incident involving one of his men.

BREAK-IN AT PEARL HARBOR

After he recuperated for six months, Brown's next job was as a radar consultant with Lockheed-Vega. He later left the California aircraft corporation, moved to Hawaii, and was a consultant at the navy yard at Pearl Harbor. An old friend who was teaching calculus there had opened some doors, and in 1945 Brown demonstrated his latest flying tethered discs to a top military officer—Admiral Arthur W. Radford, commander in chief for the U.S. Pacific Fleet, who later became joint chief of staff for President Dwight Eisenhower.

Brown was treated with respect because of who he was, but again no one signed up to help investigate his discovery. His colleagues in the navy treated it lightly because it was anomalous.

When he returned to his room after the Pearl Harbor demonstration, however, the room had been broken into and his notebooks were gone. A day or so later, as Reynolds remembers Brown's account of the incident, "They came to him and said, 'We have your work; you'll get it back.' A couple of days later they gave him back his books and said, 'We're not interested.'"

Why? Brown was given the answer that the effect was a result of ion propulsion, or electric wind, and therefore could not be used in a vacuum such as outer space. The Earth's atmosphere can be rich in ions (electrically charged particles), but a vacuum is not.

He was disgruntled, but not stopped. Later a study funded by a French government agency would prove the effect was not caused by "electric wind." But even before that, Brown knew that it would take an electric hurricane to create the lifting force he saw in his experiments.

Project Winterhaven was his own effort to further electrogravitic research. He began the project in 1952 in Cleveland, Ohio. Although he demonstrated two-foot-diameter disk-shaped transducers that reached a speed of seventeen feet per second when electrically energized, he was again met with lack of interest. Alone in his enthusiasm, he watched the craft fly in a twenty-foot-diameter circle around a pole. According to the known laws of science, this should not be happening. And he went on to make other spectacular demonstrations.

When La Societe Nationale de Construction Aeronautique Sud Quest (SNCASO) in France offered him funding, he went to France and built better devices and had them properly tested as well. Those tests convinced his backers that it could be a feasible drive system for outer space, he told Reynolds. SNCASO merged with Sud Est in 1956 and funding was cut, so Brown had to return to the United States.

Brown was eager to show the French documentation to all those officials who had raised the wall of indifference in the past. But after his discouraging visit to Washington, D.C., in 1956 and what felt like a put-down from Rickover, he apparently decided, "If the military isn't interested, the aerospace companies might be."

Friends say it did not occur to Brown to ask if the defense industry was already working on electrogravitics, unknown to him. In 1953 he had flown saucer-shaped devices three feet in diameter in a demonstration for some air force officials and men from major aerospace companies. Energized with high voltage, they whizzed around the fifty-foot-diameter course so fast that the reports of the test were stamped "classified."

Independent researcher Paul LaViolette, Ph.D., traces the path that these impressive results led to—toward the Pentagon, the military hub of the United States. In the early 1990s LaViolette wrote, "A recently declassified Air Force intelligence report indicates that by September of 1954 the Pentagon had launched a program to develop a manned anti-gravity craft of the sort suggested in Project Winterhaven."

Meanwhile, Brown went practically door-to-door in Los Angeles to try to rouse some interest in his work. One day he returned to his laboratory to find it had been broken into and many of his belongings were missing.

CHARACTER ASSASSINATION

Then the nasty rumors started. They were the type of rumors that can discredit a man's character, upset his wife and children, and overall cause deep distress to a sensitive man.

Another tragedy in Brown's life was the sudden death of his friend and helpful supporter Agnew Bahnson, who funded him to do anti-gravity research and development beginning in 1957 in North Carolina. Did they make too much progress? In 1964 Bahnson, an experienced pilot, mysteriously flew into electric wires and crashed. Bahnson's heirs dissolved the project.

The authors of the book *The Philadelphia Experiment* wrote that in spite of his numerous patents and demonstrations given to governmental and corporate groups, success eluded Brown. "Such interest as he was able to generate seemed to melt away almost as fast as it developed— almost as if someone . . . was working against him."

Today's researchers looking at Brown's life have noticed that he went into semiretirement sometime in the 1960s. Thomas Valone of Washington, D.C., who compiled a book on Brown's work in 1994, speculates that the work was classified and Brown was bought off or somehow persuaded to stop promoting electrogravitics. Valone told the April 1994 meeting in Philadelphia that LaViolette's detective work sheds new light on what happened to Brown in the 1950s. The speculation of these scientists is that "this project was taken over by the military, worked on for 40 years, and we now have a craft that's flying around." Valone speculates that Brown was debriefed and told what he could talk about.

From the later 1960s to 1985, Brown turned his attention to other related research. He mainly did basic research to try to understand strange effects he saw. As did fellow inventor T. H. Moray, Brown had decided that waves coming from outer space are not only detectable on Earth, but they also build up a charge in a properly built device. Instead of making increasingly complex devices, however, Brown, toward the end of his life (this was in the 1980s), was getting a charge—voltage to be exact—out of rocks and sand. It was all in search for answers.

If his work had been accepted instead of suppressed by seeming disinterest, he would be known to science students. His work would fill more than one science book; an encyclopedia set could easily be filled with Brown's experiments and discoveries.

For example, his childhood fascination with the singing wires led him to investigate how to modulate ionized air like that which had carried the high-voltage current. Could this be used for high-fidelity sound systems? Eventually he did invent rich-sounding Ion Plasma Speakers, which incidentally had a built-in "fac"—a cool breeze of health-enhancing negative ions. Would this discovery have been commercialized if his main interest, electrogravitics, had not been suppressed by ignorance or been co-opted?

He searched for better dielectrics, endlessly trying new combinations. (A dielectric is any material that opposes the flow of electric current while at the same time having the ability to store electrical energy.) This search led him to study, when working with Bahnson, the lighter-than-air fine sand in certain dry riverbeds that could be used to make advanced materials.

The anomalous sands were first discovered by his hero Charles Brush early in the last century. Brush also found that certain materials fell slower in a vacuum chamber than others. He called it gravitational retardation and said they were slightly more interactive with gravity. These materials also spontaneously demonstrated heat. Brush believed that the "etheric gravitational wave" interacted with some materials more richly than with others. Brush's findings were swept under the rug of the science establishment.

Brown followed his idol's lead and did basic research in a number of areas, including gravito-electrics—how neutrinos or gravitons or whatever they are, are converted into electricity. This led him to conduct experiments in various locations, from the ocean to the bottom of the Berkeley mine shaft.

When Reynolds became interested in Brown's work in the last five years of the inventor's life, Brown was able to do the work he loved the most—petrovoltaics. No one else was putting electrodes on rocks to measure the minute voltages of electricity that the rocks somehow soaked up from the cosmos. Brown and Reynolds made artificial rocks to see what various materials could do and how long they would put out a charge.

Their efforts in a number of areas led toward what they called a "forever ready battery"—a penny-sized piece of rock that put out a tiny amount of voltage indefinitely because they had learned how to "soup up" the effect. After Brown died, Reynolds carried on the research until funding ran out. He estimated that it would have taken up to $10 million of advanced molecular engineering research to take the discovery to another stage of development. The high-power version of the battery remains on paper—only theory until developed further.

This discovery alone should have put Brown into science history books. In all his years of experiments with the periodic variations in the strip-chart recordings of the output from the materials, he found that the patterns had a relationship to the position of the stars. And orientation toward the center of the universe seemed to make a difference too. This resulted in further unconventional thinking on Brown's part, which only made him more of an outcast in the world of sanctioned science.

While he was coming up with the cosmic findings, the military researchers had a different agenda. One of the reports dug up by LaViolette came from a London think tank called Aviation Studies International Ltd. In 1956 the think tank wrote a classified "confidential" survey of work done in electrogravitics. LaViolette says the only original copy of the document, called *Report 13*, was found in the stacks at the Wright-Patterson Air Force Base technical library in Dayton, Ohio. It is not listed in the library's computer.

Excerpts from *Report 13* paint a picture of heavy secrecy. A 1954 segment says that the infant science of electrogravitation may have been a field where not only the methods are secret, but also the ideas themselves are a secret. "Nothing therefore can be discussed freely at the moment." A further report predicted bluntly that electrogravitics, like other advanced sciences, would be developed as a weapon.

A couple of months later, another now-declassified Aviation Studies report said it looked like the Pentagon was ready to sponsor electrogravitic propulsion devices and that the first disc should be finished by 1960. The report anticipated that it would take the decade of the '60s to develop it properly, "even though some combat things might be available ten years from now." Defense contractors began to line up, as well as universities who get grants from the U.S. Department of Defense.

After he came across *Report 13*, LaViolette put his knowledge of physics to work and began to piece together a picture of what may have happened in the previous thirty years. It includes "black projects"— work that the military deems to be so top secret that it is kept strictly under wraps, so much so that even Congress does not get reports about its funding.

Figure 5.2. B-2 stealth bomber

A breakthrough in LaViolette's quest for the pieces of the picture came when a few establishment scientists gave out tidbits of formerly secret information about a "black funding" project—the B-2 bomber (also known as the stealth bomber).* (The B-2A was then described as the world's most expensive aircraft at $1.2 billion.) Their description of the B-2 gave LaViolette and others a number of clues about the bomber—softening of the sonic boom as Brown had talked about in the 1950s, a dielectric flying wing, a charged leading edge, ions dumped into the exhaust stream, and other clues. The B-2 seems to

---

*Richard Boylan claims the B-2 stealth bomber operates largely by using antigravity technology for its propulsion system, but this information has never been disclosed to the general public, which believe the bomber relies on conventional fuel sources. See his article "B-2 Stealth Bomber as Antigravity Craft," available online at www.drboylan.com/waregrv2.html.

be a culmination of many of Brown's observations made more than forty years ago.

Brown fought an uphill battle all his adult life, at great cost to himself and to family life. His cause included getting the science of advanced propulsion out into public domain, not hidden behind the Secrecy Act of 1951 and a wall of classified documents. He died feeling that he had lost the battle.

PART 2

Infinite Energy

A NEW SCIENCE FOR A POLLUTION-FREE WORLD

The true history of human civilization will begin with the end of energy shortage.

HARTMUT MULLER

Free energy will promulgate a forward leap in human progress akin to the discovery of fire. It will bring the dawn of an entirely new civilization—one based on freedom and abundance.

STERLING D. ALLAN

Ere many generations pass, our machinery will be driven by power obtainable at any point in the universe. . . . Is this energy static or kinetic? If static, our hopes are in vain; if kinetic—and this we know it is, for certain—then it is a mere question of time when men will succeed in attaching their machinery to the very wheelwork of nature.

NIKOLA TESLA

My perception that we could replace the polluted, depleted fossil fuel economy soon with a free energy era is not a new one. It has been with us for more than a century since the time of Nikola Tesla. It seems incredible that nobody in power or with substantial money has embraced this possibility or has taken sufficient time to look at it.

BRIAN O'LEARY

Imagine a machine that sits in a cupboard in your home producing electricity. It requires no maintenance, and—if connected to your fuse board—will produce all the energy you need to run your home. After you have bought it you need never pay for any further electricity. It also has other, wiser, advantages: unlike current energy technologies it produces no pollution (particularly carbon dioxide) and no waste products, and does not contribute to any other negative environmental impact.

KEITH TUTT

A planet is at stake. If you have the resources to launch a new energy era, do it, and don't be so concerned about what percentage your return-on-investment is.

JOEL GARBON

This vacuum energy density is calculated to be so large that the intrinsic energy contained within the volume of a single hydrogen atom is about one trillion times larger than that stored in all the physical mass of all the planets plus all the stars in the cosmos out to the present limits of detection, a radius of 20 billion light-years. . . . This makes the energy stored in physical matter a mere whisper compared to that stored in the vacuum. Uncovering the secrets of the vacuum is obviously a very important part of humankind's future!

WILLIAM A. TILLER

After the third tribulation, a new source of energy will be discovered that taps the earth's magnetic field.

HOPI PROPHECY

6

THE SUSTAINABLE TECHNOLOGY SOLUTION REVOLUTION

A UNIVERSAL APPEAL

Brian O'Leary, Ph.D.

NOTE: Text of address to the Exemplar Zero Initiative launch, United Nations, New York City, October 10, 2010. The author passed away on July 29, 2011.

I am giving this talk as if it were my last. I join the Earth and her abundant but dying life in an eleventh-hour appeal to stop the attack on all of us by human greed and aggression.

Like many of my friends and relatives, I grew up believing in the American dream. In my first thirty years, spanning the 1940s, 1950s, and 1960s, I believed that if someone had a better idea to benefit humanity and the environment, it would be responsibly researched and implemented with the support of a benevolent government and corporate system eager to help, to improve things for us all.

And so as a youth I jumped at the opportunity to serve as an Eagle Scout, Apollo astronaut, and Cornell professor. I was motivated to contribute my talents to national and global goals, only faintly aware of the growing human attack on the environment.

How naive I was. After all these years I realize more than ever that unless we quickly find ways to innovate ourselves out of the mess we have created, the world will continue to be mismanaged to the point of extinction of our own species and many others. Energy production continues to be controlled by corporations motivated solely by profit for their shareholders rather than the Earth. As a result, the Earth is becoming irreversibly polluted by the extraction and burning of dirty resources including oil and coal while other non-contaminating options are ignored or suppressed. We ignore innovation at our risk and peril.

I am now appealing to those of you with resources to team up with those of us with specialized knowledge to bring forward commonsense, cooperative, ethical, clean breakthrough technologies—or else our civilization and most of nature as we know it are doomed.

Earlier this year I turned seventy, and my life ever since has been filled with challenge and opportunity, which also reflects the condition of our world. In August alone I lost a sister and suffered a major heart attack that landed me in an intensive care unit in a hospital. This year feels like an initiation into true elderhood, with its possibly imminent physical mortality along with a heightened sense of maturity and responsibility.

CRISIS AND OPPORTUNITY

Since 1940 we have globally increased our energy use five times, our water use four times, our population three times, and we have more than half depleted our oil, freshwater, topsoil, and wood resources. I think we can all agree that we are on a disastrous course unless we change our ways.

In 2010 alone, we are witnessing Jim Garrison's "climate shock" of extreme weather events, the biggest oil spill ever, dying oceans and rain forests, melting ice caps and glaciers, increasing military and "security" expenditures, and a corrupt financial system—all controlled by corporations and governments that have yet to demonstrate they really care about the environment.

We must change this!

I believe technology can change our course dramatically, but we must overcome the resistance of vested interests so that we can allow technology the opportunity to accomplish this. Over the past twenty years, I have seen numerous demonstrations of concepts of clean breakthrough energy, water purification, and ecosystem restorations that appear to be miraculous, but are in fact very real and practical. Hard as it may be to accept, these miraculous inventions have been denied and suppressed by our mainstream culture. We have been silenced by the tyranny of vested interests.

How did we get into this mess in the first place?

According to Daniel Quinn in his historically prescient novel *Ishmael,* civilization originated and became corrupted about ten thousand years ago in what was formerly Mesopotamia when agriculture was discovered and organizations began to spread with the seizure of land and the exploitation of resources. And so began the new culture of the "Takers." Through his main character, Ishmael, Quinn wrote: "The disaster occurred ten thousand years ago when the people of the [emerging] culture said, 'We're as wise as the gods and can rule the world as well as they.' When they took into their own hands the power of life and death over the world, their doom was assured. . . . Takers will never give up their tyranny over the world, no matter how bad things get . . . they've always believed that what they were doing was right—and therefore to be done at any cost whatever. . . . Everyone has to be forced to live like the Takers, because the Takers had the one right way."[1]

And thus the indigenous cultures were forced to become the Leavers, as they retreated from Mother Culture. Most of the original cultures have sadly been exterminated by the Taker aggressors.

More recently, the oil-propelled Industrial Revolution has only exacerbated the aggression. The Industrial Revolution became the Agricultural Revolution on steroids.

Only 150 years ago, oil was discovered, marking a second epochal change in human history. The discovery of oil has led to the great riches of a new industrial civilization fueled by petroleum, much of which can be found in and within a few hundred kilometers of Kuwait.

This oil-rich land provided the first fertility after the global ice age. Ten millennia later, it became a principal provider of the abundant energy needed to grow our cities and industries into unimaginably powerful and now world-threatening proportions.

Through the years, the human hierarchies and empires grew to enormous and unwieldy sizes, and the wealth of the few began to flourish during this past century, in large part from oil drilling and consumption.

We now need a third revolution, a sustainable technology solution revolution, and may it begin to thrive once again in the cradle of civilization. Now, in the year 2010, we sit poised on either the verge of extinction or else on an unprecedented revolution in innovation and consciousness that could just give us the chance we need for our survival—*if* this can be ethically, responsibly, and consciously implemented. Call the new revolution whatever you'd like.

I often call the coming paradigm shift a "turquoise revolution," symbolizing a newly unified and healed Earth with pure oceans, land, and air.[2] Turquoise combines the best of Gunter Pauli's nature-friendly blue economy,[3] the prospects of breakthrough clean energy, and the green thinking of the deep ecology movement.

Most of us can agree that the Agricultural and Industrial Revolutions have created both great civilizations and great riches alongside great poverty, the inequalities of which contribute immeasurably to the destruction of our precious environment.

We need to undo ten thousand years of tyranny and transform our Taker-dominated culture into a Giver-dominated culture. But where are the Givers, where can we find those of us who really care about our future? How can the few Takers with a conscience transform themselves to Givers?

The first step in such a process is to let go of our fears and become aware of the magnificent things that are possible. When we do that we begin to realize that there are many innovative solutions available. The problem is that they seem so elusive. A juggernaut of industrial tyranny is cutting us off from these solutions, and the majority of people go along with the charade. Whether it's our land, water, food, oil, wood, metals, or money, the grab is on.

But there's a missing piece in this puzzle. Those who have correctly pointed out our dilemma almost universally refuse to acknowledge the true possibilities that lie ahead. The missing piece is that we can uncover the truth about innovative technologies and then convince key people to make the wise choice of truly sustainable technologies free of vested interests and promotional biases.

Ironically, those of us who should know better—the scientists, environmentalists, political progressives, academics, and media—still join the herd of Takers in not even acknowledging the *possibility* that our energy could be clean, cheap, safe, abundant, and decentralized. So we become unwitting allies with the powerful elite and miss the greatest opportunity we could possibly imagine to resolve our dilemma.

In a recent published essay [*Infinite Energy* magazine, September/October 2010], I urge the scientists, environmentalists, and progressives among us to be among the first to embrace the possibility of producing free energy and other breakthrough clean technologies. I consider myself a member of all three communities, whose critiques are spot-on but whose solutions are too little, too late. Buckminster Fuller said: "There is only one revolution tolerable to all men, all societies, all political systems: revolution by design and invention."

So far, I have received little response from my scientific, environmental, and progressive colleagues to my pleas to consider the possibility of an energy solution revolution. Perhaps now, however, as more and more of us become aware of the environmental devastation wrought by BP's Gulf of Mexico ecodisaster, Pakistan's floods, and Russia's fires, we may reach a tipping point in our efforts to awaken to deeper solutions.

BREAKTHROUGH GREEN INNOVATION

There's a lot of talk about green technology solving our climate and other environmental problems. But what is green technology? Is it the water- and fuel-intensive green revolution of monocultures wracked by fertilizers and pesticides? Biofuel plantations that destroy our forests and soil and take food from the mouths of the hungry? Nuclear power, fossil fuel power, and hydropower plants involving massive grid systems? Geoengineering? Even the seemingly renewable but materials-intensive windmill and solar farms producing intermittent electricity and hydrogen to fuel our fleets?

My radical viewpoint is that none of the above measures will prove to be nearly enough to solve our collective dilemma after we factor in the full life-cycle environmental costs.

To cocreate a truly sustainable future, we will need to design new technologies, especially energy technologies, from the ground up. Fortunately, there exist hundreds of proofs of concepts of new energy devices, ranging from energy from the vacuum (zero point), cold fusion, and special hydrogen and water technologies, and energy from the thermal environment.[4]

All these approaches deserve our closest consideration, research, and development, but sadly the powers-that-be have suppressed these initiatives because free energy threatens the continuing viability of existing energy approaches. Consideration of free energy has been quashed from all discussion, either because of disbelief or fear of disrupting the status quo. "Status quo is a multidimensional tapestry of what has been and will never be again. And is, ipso facto, no longer existent." Not that we should drop everything we're doing in the near term to improve efforts toward sustainability. I'm a dedicated advocate of small-is-beautiful and the brilliant work of Gunter Pauli, John Todd, and others in sustainable biosystems that work with, rather than against, nature. Ongoing research and development of some of these bridge innovations will become absolutely essential in the choices we need to make.

Whether it's an oil, mining, logging, water, agribusiness, or pharmaceutical company coming into the Amazon or Arctic, a missionary or bank incursion into indigenous lands, a gringo coming to the South to become a land developer for short-term profit, the suppression and hijacking of miracle breakthroughs like authentic cures for cancer or research on free energy, political or financial manipulation of vulnerable nations with resources, or a coup d'état, assassination, or military attack, the story is always the same: the Takers run the show. We've been "hoodwinked" says whistleblower-author John Perkins.[5] Almost all of us are unaware of how serious the situation has become and how ignorant we are of new possibilities.

The key to understanding the roots of our destructive way is to understand the practice of economics. The "dismal science" of Thomas Carlyle has incorporated arbitrary standards that favor the rich and externalize environmental costs. Under these rules the BP oil explosion in the Gulf of Mexico could be considered an economic success (more jobs, more profits for some, higher GDP) but not accounted for as an environmental catastrophe.

I am deeply saddened by our unwillingness to halt our predatory environmental destruction. I often feel I need an ecopsychologist to plumb the depths of my own grief about what's happening so quickly to our planet.[6]

I often feel angry, depressed, betrayed, guilty, and fearful about what our species is doing to itself and its surroundings—all because of the bankrupt values of current corporate and political policies. The industrial age is crashing down all around us, and no publicly understood strategy for our future presently exists. Most of us don't even know it's happening.

But, just as the agricultural age replaced the Neolithic age and the industrial age replaced the agricultural age, another age, perhaps the age of wisdom, is urgently required to replace our current age of greed if we are to survive. We cannot go on like this or allow the ecocide to continue. It's time for those of us who are consciously evolving to take a stand on behalf of life, to say "enough!" and demand real change and action—not in ten or twenty years, but here and now.

Waiting in the wings is a cadre of enlightened, practical visionaries ready to move forward with countless clean innovations. We should begin to provide sanctuary for true innovators so that they can conduct their research and development in peace and help us cocreate a much better world. These unsung pioneers deserve our financial and emotional support.

A change in consciousness has become absolutely necessary. Our greatest obstacles are social and political, not technical. There are many good ideas out there waiting for their opportunity, ideas that have sometimes been violently suppressed at the technology-development stage because of the greed, avarice, and addiction to power of the vested interests. It is time for the many to receive the benefits of abundant sources of energy rather than the few at the expense of the many.

It's time to ask how we, as a civilization, can best embrace sustainability while minimizing disruptions during the transition period. We all need to become educated about what's possible and then create teams of passionate ecoscientists and engineers to fulfill our collective mandate for clean energy, clean water, sustainable agriculture, forestry and oceans, and the protection of the natural world and its biodiversity.

FIVE TECHNOLOGY SOLUTION REVOLUTIONS

"Necessity is the mother of invention." We need new clean technologies to match the boldness of our goal to cocreate a truly sustainable future. We can develop them if we just give them a chance. I foresee at least the following revolutions coming out of the work being proposed at the launch of initiatives like the Exemplar Zero Initiative.[7]

An Energy Solution Revolution

Time and again, I've seen demonstrations of proofs-of-concept of numerous breakthrough energy technologies, many of which, if properly developed, could solve the energy crisis very soon and eliminate most causes of climate change and rampant pollution.[8] These technologies include: (1) energy from the vacuum extracted by electromagnetic devices (sometimes called zero point, etheric, space, free, over-unity energy), (2) cold fusion or low-temperature nonradioactive nuclear reactions, (3) advanced hydrogen and water chemistries, and (4) energy from the thermal environment. Yet, as a culture, we've repeatedly rejected this possibility by not supporting the research; in fact, this research has been violently suppressed.

This situation is the most bizarre conundrum of our time as we continue to neglect even thinking about something that could save us from our collective folly. Consciously or unconsciously, the controlling elite are committing genocide and ecocide, while the rest of us reject the most potent and effective means of preventing these unpardonable acts.

We must now demand research, development, and assessments of many technologies in order for us to be able to launch a viable program that truly benefits all humanity. According to Buckminster Fuller, "People should think things out fresh and not just accept conventional terms and the conventional way of doing things."

This new energy program should mirror John Kennedy's vision to place a man on the moon. Where are the John Kennedys of today? Perhaps some of you may feel the stirrings in your souls for greatness and bold action that our time requires.

Take a moment and imagine a world of free energy—no more extraction and burning of dirty fuels, no more grid systems, vastly reduced air, water, and land pollution, no more murderous trillion-dollar wars for oil. Imagine holding a ten-kilowatt power pack in your hand that could power your home or transport you to work. Is this an impossible dream? Actually, from what I've seen and learned over many years, this future is highly likely if we will only give it a chance.

Imagine in your daily lives no longer having to pay for an electric or heating or cooling bill or for "filling 'er up" when you want to drive or fly anywhere or ship anything anywhere. Imagine having unlimited, safe, fresh drinking water worldwide from the cheap desalination of seawater. Imagine a world without big power plants and huge dams but rather an elegant decentralized system of distributed power reducing our cost of living by 30 percent or more. Imagine ending poverty worldwide from this one revolutionary development—free energy—and thus fulfilling the United Nations' Millennium Development Goals. Again, some words of wisdom from Buckminster Fuller: "I am glad I have lived long enough to see this! It is simply wonderful! I hope and pray that you will live long enough to see the principle upon which this marvelous artifact is based become the new energy source for all the passengers on Spaceship Earth."[9]

It's logical that energy companies earning hundreds of millions of dollars a day would try to maintain their income streams for as long as possible and wring out every drop of oil before changing course. So it's up to the most enlightened leaders of nations and companies as well as the rest of us to awaken to the possibility of a clean and sustainable future through free energy and other breakthrough technologies. We must compel the largest companies and countries to make ethical and responsible choices that are no longer predicated on industrial self-interest but rather in the interest of life itself. Bucky Fuller again: "We are not going to be able to operate our Spaceship Earth successfully nor for much longer unless we see it as a whole spaceship and our fate as common. It has to be everybody or nobody."

A Water Solution Revolution

Water is the wondrous substance on which we each rely for our survival. Sometimes called the oil of the twenty-first century, water is declining rapidly in both quantity and quality. Instead of keeping water pure to ensure our health and the health of the planet, we have treated our bodies and waterways as toxic dumping grounds.

Pure water is the bridge to higher consciousness within and around us.[10] Recent research suggests that we can heal ourselves and our planet through the medium of water (which could be considered an essential *blue* element to our *green* thinking). As hard as it may be for many of us to imagine, much less believe, researchers, including the late Rustrum Roy, Marcel Vogel, William Tiller, Patrick Flanagan, and Masaru Emoto, have demonstrated that we can purify and energize water, both inside and outside ourselves, through positive intention, visualization, and the vortex science of Viktor Schauberger and others.

The wondrous relationship between consciousness and water suggests that we are far more intimately connected to our world and one another than we ever dreamt of in the old mechanistic paradigm. This is the world we've inherited, the world we can reinherit in a new way, if we can openly discuss what it will take to restore what we have so badly abused. If we are to achieve the immediate goal of sustainability, radical innovations in design using nature as a template must become our first priority.

A Localized, Sustainable Natural Systems Agriculture/Forestry/ Water/Energy/Waste Management/Economics Revolution

Several innovations enable us to create true sustainability at a local level in balance with nature by conserving and reusing water, topsoil, energy, waste management, and ecological construction. These plus increasing the use of raw and organic foods, permaculture, herbal medicine, ecological economics, and the practice of vegetarian lifestyles all add to a sustainable mix that can strengthen communities. Ongoing research on the state of the art of these methods, independent of fossil fuels and other unsustainable practices, should become the very foundation for creating locally vibrant communities and bioregions with strength and sovereignty, while contributing to the global commons as well.

A Biosphere Solution Revolution

We not only need to sustain our local communities, we must also preserve what land, water, and air is still intact, and we need to restore damaged ecosystems worldwide. Research shows that these areas could be brought back into their former pristine state with the return of flora, fauna, rich soil, and freshwater through careful planning as applied to the area to be restored.[11] Environmental writer Eugene Linden wrote:

> Peter Raven, director of the Missouri Botanical Garden, predicts that during the next three decades man will drive an average of 100 species to extinction every day. Extinction is part of evolution, but the present rate is at least 1,000 times the pace that has prevailed since prehistory.
>
> Even the mass extinctions 65 million years ago that killed off the dinosaurs and countless other species did not significantly affect flowering plants, according to Harvard biologist E. O. Wilson. But these plant species are disappearing now, and people, not comets or volcanoes, are the angels of destruction. Moreover, the Earth is suffering the decline of entire ecosystems—the nurseries of new life forms. For that reason, Wilson deems this crisis the "death of birth." British ecologist Norman Myers has called it the "greatest single setback to life's abundance and diversity since the first flickerings of life almost 4 billion years ago."[12]

Restoration of the biosphere is a third mission-critical goal. We will need to leave undeveloped lands alone, use state-of-the art methods in restoring damaged ecosystems to their former pristine state, and develop local organic agricultural methods, standards, and certifications that are truly sustainable.

We could create an Earth Corps to restore ecosystems and provide jobs to those who lose them as a result of the transition to a new culture, the funding for which could derive from reduced military budgets and from taxes on pollution. And we could employ Colonel Jim Channon's vision for a First Earth Battalion that uses the army to restore the land, the navy to restore the oceans, and the air force to restore the air.

Ongoing Protected Research and Development Centers

We will need to have cooperative international networks of innovators and local teams working at protected centers to research, assess, develop, test, and evaluate the long list of sustainable technologies that can replace our toxic manufacturing waste streams.

So, what if some breakthrough technologies don't work or turn out to have negative environmental impact? No problem, we just continue researching others. There are literally hundreds of basic ideas and thousands of pathways to achieve success. Again quoting Fuller: "There is no such thing as a failed experiment, only experiments with unexpected outcomes. I only learn what to do when I have failures." You don't even have to believe in free energy to embrace its possibility!

That's what R and D is all about—to invest in many parallel technologies at various stages of development in an iterative, integrated effort to guide us to the best choices based on health and environmental friendliness and not on immediate profits or promotional biases toward particular systems.

New Design Imperative

Imagine a decision-making process that places people and the planet before profits and asks if a potential project enhances life, and only if the answer to that question is yes does the project proceed to evaluate the economics. If it harms people or the environment, it is dead in its tracks. Contrast that with the normal way of evaluating a project that simply says, "This will make billions, let's go!" We live with the many negative consequences of not thinking about consequences. A new life-enhancing design imperative must become the norm.

I like Buckminster Fuller's expanded definition of a designer: "A designer is an emerging synthesis of artist, inventor, mechanic, objective economist and evolutionary strategist" and "Design science is more than the application of engineering and technology. It is more than a plan or a design. Design science means the total responsibility and capability for development, production, and distribution—of not just a product—but a total service system on a worldwide basis."

My experience in the aerospace community has taught me that the first step in any design process is identifying the design *requirements*. Here I think most of us agree that the overriding requirement in any significant design for our future is true sustainability.

This initiative will need quality teamwork as in the Apollo program. In the long run, establishing properly managed R and D and technology assessment centers would lower the risk of prematurely picking technologies that wouldn't work in the long run, like internal combustion engines, nuclear power plants, grid systems, biofuel plantations, huge dams, and other approaches that someday will become obsolete.

Can we abuse free energy and other breakthroughs like we have abused so many others like nuclear energy? Of course, we can. But we don't have to if we design and implement our systems wisely with true sustainability rather than profit and vested power as our central criterion and focus.

Earlier in 2010, seven of us gathered at Montesueños, the conference bed-and-breakfast retreat center my wife and I built in Ecuador, to help begin the process. Called the Global Innovation Alliance (GIA), we are beginning to focus on integrating truly sustainable technologies.

The GIA is an international nonprofit network of like-minded, competent, and visionary innovators inventing, advancing, and integrating sustainable technologies to facilitate independence from oil. We are a global organization seeking truly sustainable, exportable, and scalable solutions at fundamental levels, free of vested interests.

We seek to develop sanctuaries that would support and protect the R and D of innovative systems that are self-sustaining in energy, water, waste management, and agriculture, both near-term and longer-term. We are facilitators, integrators, and assessors of misplaced resources, ameliorating existing technologies and designing solutions that provide a generous return on investment for every passenger on Spaceship Earth.

Our mission as described by Buckminster Fuller is "the effective application of the principles of science to the conscious design of our total environment in order to help make the Earth's finite resources meet the needs of all humanity without disrupting the ecological processes of the planet."

The basket of technologies contained in the Exemplar Zero Initiative represents one beginning of what we need to do. They remind us that many cutting-edge innovations, of which the vast majority of us may be unaware, can change the world very soon. I also envision that once developed, these innovations will be joined by many other technologies that can be integrated and become mature with time as the R-and-D process and assessments take place.

Here I must interject a note of caution. For over one hundred years, since the time of Nikola Tesla, free energy and other disruptive technologies have been consistently and sometimes violently suppressed by a system that favors the powerful, who gladly support research on weapons and dirty energy. The brilliant scientist-author-inventor David Yurth put the situation this way: "The most dangerous of all human undertakings is the practice of science without a conscience. If we do not care for all living things on the planet, we cannot pretend to be genuinely motivated to problem solving."[13]

And the converse is also true: the most dangerous of all human undertakings at a *personal* level is the practice of science *with* a conscience. Whether it's the corrupt patent system, the unfavorable investment attitudes about putting money on the line for perceived "risky" development, inventor naiveté about how the real world works, the overall ridicule foisted on these initiatives by skeptical but uninformed scientists, or the theft or burial of the intellectual property by large corporations or governments beholden to the status quo, history has consistently shown that almost every time a promising but disruptive breakthrough technology has made progress, the effort has been sabotaged.[14] In the case of free energy technologies, the suppression has been complete.

So any radical, new sustainable technology development will need to have the unanimous, transparent, and cooperative support of new creative partnerships between governments, entrepreneurs, and protected research teams. "When individuals join in a cooperative venture, the power generated far exceeds what they could have accomplished acting individually."

Likewise, any aspiring Bill Gates for free energy will need to have his own control and money agendas filtered out of the planning. Huge growth, profits, and wealth should not be a part of our quest. Fuller once again: "I learned very early and painfully that you have to decide at the outset whether you are trying to make money or to make sense, as they are mutually exclusive."

Truly sustainable technology development needs to be shared with the world, and yet the intellectual property of the inventor needs to be protected and rewarded. This blend is a tall order in a capitalistic world where moneyed interests would like to breed more money—for themselves. Open sourcing, government support, and private donations motivated more by altruism than profit will need to be the new standard we all should strive for.

7

POWER FOR THE
PEOPLE—FROM WATER

Jeane Manning

We inhabitants of this planet have made huge mistakes. The good news is that we could turn the situation around by cooperating—with each other and with the rest of the natural world. For perspective, we could begin by viewing the scene from a distance.

Earth is a blue jewel in our solar system—the water-rich planet. A closer view reveals abuse of that priceless water. Oil blackens ocean beaches. Aging barrels of radioactive waste leak into ground and ocean waters. Coal mining blasts mountaintop springs and water-enlivening forests into oblivion. Hydro dams create stagnant lakes on previously healthy rivers, and methane mining harms other streams. Supplies of drinking water shrink, and much of this is caused by the dominant energy technologies.

Wiser beings viewing Earth might wonder, "Will the inhabitants of this planet use their emerging crises as springboards for cooperative actions? Will they unite to shift to energy technologies that are in harmony with nature? Will they employ the harmonious technologies to restore watersheds, cleanse rivers and oceans, and ultimately create a higher civilization?"

It's possible. Authors in this anthology point out various paths the human family can take toward environmental stewardship and a more caring society. Some contributors see the need for a new source of energy beyond the standard alternatives. Some find that individual inventors and independent research groups around our world *have* made the breakthroughs that could replace coal, oil, and uranium. This chapter introduces water-related breakthroughs. They range in size from Blue Energy Canada's hydroturbine tidal-power bridges down to the injecting of water vapor into a car engine. Some innovations swirl water into the form of an inward-spiraling vortex—nature's choice for an energy-tapping motion—and some similarly involve imploding bubbles in water to release powerful energies. A few types of the inventions could clean up polluted water at the same time as providing electric power.

Humankind knows how to harness fire, wind, sun, and the strongest forces in nature. Will the next leap forward use the power of water? That question is posed by the BlackLight Power website (www .blacklightpower.com). It describes a novel hydrogen-related energy that's cheap, abundant, and generates no pollution or greenhouse gases. A video on the site says our prosperity is limited only by our ambition. We can keep our world, oceans, rivers, and the air we breathe clean by using "energy made right here at home—or anywhere in the world." This chapter concludes with good news about that BlackLight process and about global cooperation among new-energy activists, but first will look at other water-related discoveries down through history.

DAWNING OF THE AGE OF AQUA

The urge to solve humankind's energy-generating problems is felt by innovators around the planet. Thousands of videos and forums on the Internet share research and homegrown experiments. Before the Internet and its open sourcing of knowledge, lone inventors tried to single-handedly save us from our dependence on dirty fuels. And long before hippies saw the dawning of an Age of Aquarius, a few individuals discovered that water is a key to independence.

James Robey of Kentucky compiled a history titled *Water Car: How to Turn Water into Fuel*. It begins with the Swiss-German physician

Paracelsus about five hundred years ago noticing that a flammable gas formed when iron reacted with a certain acid added to water. In his era the pace of discoveries was slow. Centuries passed until an eighteenth-century British scientist had the resources and patience to isolate what he called "inflammable air" and prove it is a separate element. When the gas burned, water formed again. He figured out that water is made from two parts of hydrogen and one of oxygen. A British team was also first to perform electrolysis of water. They used a new invention, the battery, passing its electric current through water to produce hydrogen and oxygen, which appear as tiny bubbles rising out of the water.

Isaac de Rivaz of Switzerland, first to patent an internal combustion engine, powered his clunky vehicle on hydrogen from water. It took the wrong path; fuel was not created on demand, but instead hydrogen was dangerously stored in a tank before being released by hand into a five-inch-diameter cylinder and sparked. His 1805 car lurched forward fifteen feet every five seconds. There was room for improvement, but the car's exhaust was clean.

Michael Faraday was a British scientist who said, "Nothing is too wonderful to be true." Unfortunately some of today's experts use his fame dogmatically to justify a narrower viewpoint. In 1834 Faraday's experiments proved a certain limit to the efficiency of electrolysis. The amount of hydrogen and oxygen output from Faraday-type electrolysis is a set proportion; nearly four kilowatt-hours of electricity is needed to produce one cubic meter of hydrogen. Twenty-first-century experts invoke "Faraday's law" to shut the door on maverick experimenters' claims of superefficient electrolysis.

Faraday's formula does hold true for the usual electrolysis experiment using "brute force" direct current. You can't run a generator on water by burning the hydrogen gas put out by a standard electrolyzer (the apparatus that uses electricity to break apart water). It takes too much electricity to produce the gas, so burning won't ever produce enough heat to be converted to that much electrical energy. Such a system couldn't power a generator or vehicle. However, not everyone builds things the usual way. And not everyone breaks up water with the sledgehammer approach of a heavy flow of electrical current. Instead, approaches that require less electricity can be used to separate hydrogen atoms from water molecules. Those who are doing "non-Faraday electrolysis" range from backyard tinkerers to highly educated maverick scientists who may have an understanding of atoms that is advanced beyond mainstream science.

In 1872 acoustic researcher John Worrell Keely of Philadelphia (please see chapter 2) reportedly demonstrated the use of multiple sound vibrations to break water apart. Could a combination of tones from three tuning forks produce music that causes hydrogen atoms to dance their way out of the water molecule? Keely said he had indeed found the resonant frequency of water. Many of Keely's discoveries, prototypes, and papers disappeared after his death.

In Dallas, Texas, more than a century after Faraday, Henry Garrett's invention turned water into fuel on demand under a car's hood. He had previously invented the emergency dispatch radio, which he gave to Dallas without cost, and created his city's first automatic traffic signal. Garrett and his son Charles worked on their electrolytic carburetor for eight years before Charles patented it in 1935. Newspaper articles said it substituted water for gasoline and that Garrett claimed cooler motor operation, instantaneous starting in any weather, elimination of fire hazards, and full power and speed. Was Garrett a con man, as debunkers claim? There's no evidence for that charge, says Robey.

Robey dug into Dallas newspapers and other archives and found Garrett to have been a gifted technician and upstanding public servant who donated much time and many of his assets to help others. Robey wonders why the Garretts' invention wasn't used by any car manufacturer or offered as a retrofit device. "Maybe it was a few local oilmen who convinced them to just drop the thing . . ."

Meanwhile in Bolivia, Francisco Pacheco's fascination with an electricity-generating fish led him to an invention that produced fuel from water efficiently. The U.S. vice president at the time, Henry Wallace, took a goodwill tour of South America, met Pacheco, and saw the potential for his invention to replace gasoline diverted to fight World War II. Wallace invited him to immigrate, so Pacheco moved to New Jersey. By the time Pacheco consulted a lawyer about patenting his invention, however, Americans were getting ample supplies of gasoline. Few worried about pollution. Pacheco was advised to wait. Family and job responsibilities took his attention for nearly thirty years before he began demonstrating how he could power engines and a home with his device. It produced hydrogen as needed—by an interaction of salt water, magnesium, electricity, and carbon. He also ran a boat on seawater. In 1992 his second patent described a process that needed no electricity. Despite his many efforts to get fair publicity and to reach decision makers, his discoveries were ignored.

Andrija Puharich was a medical doctor whose technical brilliance led him to also earn a Ph.D. in physics and thirty patents on his inventions. After studying the famed inventor Nikola Tesla's findings about electrical resonance, Puharich tuned into the resonant frequencies of water molecules to loosen the bonds between hydrogen and oxygen. His friends claim he traveled thousands of miles through Mexico and the United States in a motor home powered by hydrogen split from water, created on demand in the vehicle. The legend includes a trip through a mountain pass where he melted snow water to fill his fuel tank. Puharich's brilliance didn't settle into a business groove, so his invention didn't make it to the marketplace. Instead he turned his attention to understanding the nature of consciousness.

Puharich and Pacheco are dead, but other inventors continue. In the Philippines, engineer Daniel Dingel says that in the past forty years he has converted more than a hundred cars to run on water, using seawater as the electrolytic solution. Textbooks say it is impossible to get a car battery to put out enough electrical power to do what Dingle claims to be doing. Skeptics point to slight traces of carbon found in his car's exhaust. However the traces don't prove he is using some hydrocarbon fuel. Instead, that carbon could have arrived in the air intake from smoggy, sooty city air. Old videos show him running a Toyota Corolla car with his small hydrogen reactor hooked up to its engine. Dingle was told that his country's government, deep in debt, could not support his efforts because that government has been instructed to avoid competing with the World Bank's energy interests—oil.

NEW WATER SCIENCE
FROM RUSSIA

Highly educated but underemployed Russian scientists are often ahead with future science despite financial challenges. Philipp Kanarev, Ph.D., a physicist, a distinguished professor, and the head of the Theoretical Mechanics Department at Kuban State Agrarian University, has at times done his own experiments without the benefit of basic comforts such as heat in his laboratory. Kanarev writes about how today's energy technologies cause ecosystems to deteriorate, but he adds that the deterioration can be halted by using water as an energy source.

Kanarev says his water plasma electrolysis method is the most efficient way to get cheap hydrogen from water. Unfortunately, his 1987 report about his method was kept away from the news media or public scientific journals for years afterward; the military-industrial complex in the Soviet Union classified the report. He focused at that time on using his device for purifying and disinfecting water. That focus changed after Stanley Pons and Martin Fleischmann announced in 1989 that they had achieved excess energy output during a type of electrolysis (see chapter 11, "Cold Fusion," by Edmund Storms).

That "tabletop fusion" announcement renewed research efforts behind the former Iron Curtain as well as in the West. As a result, in 1996 some of Kanarev's colleagues publicly announced the excess energy output from their plasma process, the next year they applied for a patent, then a full team of Russian scientists tested the device. In recent years it has been measured releasing up to ten times more hydrogen than ordinary electrolysis.

Various new-energy inventions mimic aspects of how nature works, in contrast to insensitive-to-nature old-energy systems. Kanarev observed that the way humankind has been creating megawatt or gigawatt energy systems is inefficient; it's uneconomical to use processes that operate continuously. Instead, he suggested copying the principle used by "the main natural motor—the heart of the man and the animals."

If a heart operates one-third of the time and rests two-thirds, its pulsations can be described as an impulse process. Taking that impulse principle into electrolysis means employing work-and-rest cycles (pulses) instead of a continuous flow of electricity. Kanarev and his colleagues found that when water molecules are shaken apart by electric pulses at frequencies that match the substance's natural resonant frequencies, the atoms' outermost electrons absorb energy from the surrounding space—the background energy of the universe. Perhaps this explains why Kanarev's "plasma electrolysis" can output ten times more hydrogen than Faraday's law allows.

Kanarev concludes that modern physics and chemistry fail to do electrolysis efficiently because textbooks are based on mistaken ideas about the orbital motion of electrons in atoms. He writes new textbooks based on emerging science that goes beyond the mainstream worldview.

WATERFALL, HURRICANE, DAM-BLAST FREE ENERGY

Meanwhile in the United States another well-credentialed and dignified physicist, Peter Graneau, also authors books about new scientific understandings and unlimited renewable energy from ordinary water. In an editorial in the May/June 2010 issue of *Infinite Energy* magazine, he says the most promising technology of a new clean-energy source is that of liberating hydrogen-bond energy from ordinary water at its normal temperatures.

Graneau's series of experiments, begun in the 1980s at Massachusetts Institute of Technology, led to his learning that anomalous energy and force can come from exploding water.

He did experiments described as water arcs, in which a high-voltage electrical spark (like a short piece of lightning) sets up mechanical tension in a column of water. The stress breaks the liquid into droplets of fog, and the liberated energy—previously involved in bonding molecules together—accelerates the droplets into a powerful explosion, reaching speeds as fast as one thousand meters per second. From a thimbleful of water, the water explodes with a force that cuts neat holes in quarter-inch-thick aluminum plate. More kinetic (motion) energy comes out of the water explosion than was put in through the electricity. Strangely the water does not get hot. It takes far less electricity to liberate abundant energy by breaking water apart that way than it does to heat water to the boiling point to create steam for driving a turbine.

Noting that people must be baffled by why the news media and the U.S. Department of Energy are silent about the promising discoveries, Graneau explained that the wall of silence probably rises from a misunderstanding of the basics of water science. Chemists know about the bonds between hydrogen molecules in water, but don't yet know that the bonds can be ruptured with mechanical tension, very efficiently, and that the freed bond energy can be captured in various ways. Graneau said, "We are very familiar with thermochemistry and electrochemistry, but mechanical chemistry is new to most of us."

Could the water arc be scaled up to a size that would rival the output of nuclear or coal-burning power stations? Perhaps not easily; Graneau figured that designing a multitude of electric-arc accelerators to power a gigawatt electric power generating station is too complex, so he looked for a more practical way to free the internal energy in water. He looked to nature—water breaking apart as it hits rocks on the bottom of a waterfall. Something anomalous happens there to accelerate a film of water.

Searching the scientific literature, he found no insight on what happens to hydrogen bonds when water crashes onto a hard surface, changes direction ninety degrees, and speeds outward along the surface increasingly faster. Whatever causes that acceleration could point toward a more continuous and massive rupture of hydrogen bonds, Graneau realized. Drag forces could be breaking the bonds in water as it flows radially from the waterfall impact, just as those forces do in a hurricane when it contacts the ocean surface and released energy self-intensifies the hurricane. A similar release may be accelerating the flow sideways from under waterfalls.

Graneau then realized that an innovative turbine system that breaks bonds between water molecules as well as converts the energy of falling water into electricity would be simpler to design than scaled-up water-arc technology. Other scientists would also have to do and share experiments about what happens when a fall of water hits a horizontal surface and sideways expansion of a thin layer of water accelerates. The data could be put to significant use. Graneau envisions that the power of that acceleration could be gathered by blowing it into a spiral blade structure of an innovatively shaped surrounding turbine. A small model of what Graneau calls a "spider turbine" was photographed for an earlier issue of *Infinite Energy*.

In the same May/June 2010 issue of *Infinite Energy* as Graneau's editorial, mechanical engineering expert Farzan Amini in Iran revealed other aspects of the power of water—pulsating helical vortex action in water and low-energy nuclear reactions (LENR, previously called cold fusion). To imagine what a helical vortex looks like, think of the DNA double-helix spiral. Amini's conclusions are reinforced by his analysis of a 2009 explosion at a hydroelectric plant in Russia. The blast resulted from hydrogen bonds having been broken, freeing hydrogen, which ignited. Multiple miniature explosions could have been ignited by friction or by hydrogen contacting a violent plasma jet or an electric field. Explosive forces were more energetic than could be handled by a defective turbine whose bolts were loosening. The spinning turbine rose up and destroyed machinery while pressurized falling water continued to flood the room.

Why should we nontechnical people care what caused an explosion that brought down a hydroelectric plant's turbine hall ceiling? The answer has to do with the potential for a better way to generate electricity, because a gigantic increase in the amount of energy occurred accidentally in one of that dam's 156-ton turbines. If such forces were deliberately created and harnessed safely in power generating stations, there would be no excuse for building coal-burning power stations—or eight-hundred-foot-high dams.

The forces could be harnessed without destroying things. The Sayano-Shushenskaya explosion in Russia (referenced above) involved vibration of faulty equipment resulting in resonance that amplified the vortex power and pulsations being created in swirling water. Jets of hydrogen plasma were then created, electrons captured, and hydrogen bonds broken. Engineers try to avoid these conditions, but Amini says hydro turbines can act as a reactor for the vorticity and LENR process. Being a respectable engineer, he avoids speculating on what dramatic energy abundance would mean for people's lives.

On a smaller scale, an engineer in the United States learned how to constructively harness a destructive force—the "water hammer" of common old household plumbing. Shock waves cause pipes to loudly rattle and knock when pressure inside suddenly changes. Industrial heating engineer James Griggs grasped such a pipe one day, felt the significant heat, and realized that the problem could be a solution—a new source of power. He developed, inside a metal drum, a rotating device whose indentations created shock waves that released heat into the churning liquid. The heat released represented at least 30 percent more energy than was used to electrically rotate the apparatus. It's not enough to make sufficient steam for turbines that make electricity, but enough to fire up the skeptics. They can't argue with success, however. The Hydrosonic Pump is sold as industrial-sized boilers that don't require any fuel to be burned in order to make steam or to heat liquids. Around the city of Atlanta, Georgia, public buildings have been using the device to save about a third of their heating costs. Customers include a fire station, a dry cleaning plant, a gymnasium, and the Atlanta Police Department.

Skeptics cite the physics law of conservation of energy, which says you always have to lose energy and can't gain more than you put into something. That law was written when the steam engine was first studied. It is always valid in a closed system in which every input can be measured, but what if the imploding bubbles in Griggs's and other people's inventions are tapping into the background energy of the universe? Kanarev, for instance, told me that his plasma electrolysis system seems to involve input from what was once called the aether. Flawed experiments early in the twentieth century were cited by physicists as disproving the existence of such a nonmaterial field, but dissident scientists, including Tesla, have shown evidence of it all along, although not in the static form that the nineteenth century worldview described.

Back in the former Soviet Union other scientists have looked at vortex processes and imploding bubbles as a better way to generate electricity. The late A. I. Koldamasov's device vibrates a mix of waters through a special material to produce heat energy in more abundance than the energy that powers the oscillator. The device was reported to put out forty kilowatts of heat energy with only two kilowatts of electrical input.

Vladimir Vysdotskii, a scientist with Kiev National Shevchenko University in the Ukraine, and colleagues are continuing with a similar invention involving the cavitation (violent caving-in or implosion) of tiny bubbles formed in fast-moving jets of water. He sees it as one of the possible ways to develop new energy technology.

The most passionate pioneer of vortex power was Austrian forester Viktor Schauberger (see chapter 3, "Viktor Schauberger" by Callum Coats). He used his knowledge of nature's energy-gathering movements and built environmentally friendly energy machines. Throughout his life Schauberger was an extraordinarily keen observer of water's ways in pristine ecosystems. Watching a babbling mountain stream, for instance, he noticed that each time it encounters a rock the water whirls and draws in air. The water breathes. In the spiraling of water, he recognized a basic movement of nature. He also noticed how turbulence plays its role in what becomes a pattern of natural self-organization—a stable, pulsating structure of water can be created out of swirling chaos. In Schauberger's view, Nature is there for us to understand and then copy.

Schauberger built technical devices aimed at imitating those spinning movements of water that create special results. He built special twisted pipes that encouraged water or air to swirl and accelerate from suction, not rely on pressure. He made a home-appliance-sized power converter using vortex power. Today's experimenters find it very difficult to reproduce his power generator. His grandson Jorg Schauberger advises researchers to develop their own ideas for a future of working with nature. The Schauberger message is "C^2—Comprehend and Copy Nature."

Today's dominant energy technologies do everything the wrong way, according to Schauberger's view. Instead of nature's creative motions—inward-spiraling, implosion- or suction-based, quiet, cooling, self-organizing vortices—the energy technologies that rule the world use pressure, explosions, heating, or burning and involve noise resulting from friction. No wonder people believe that technologies always create problems. They might consider that the *type* of technologies creates such messes and that visionaries such as Schauberger pointed in a different direction.

MORE WATER-AS-FUEL HEROES

As with Garrett and Puharich, Stanley Meyer was an American who drove a vehicle—in his case a modified dune buggy—running on water alone. He used hydrogen jolted out of water with the aid of resonance and a complex of related discoveries such as how to fracture hydrogen gas by ionization. Meyer said his process made the gas become very powerful, so a reduced amount could do more work than hydrogen from traditional electrolysis. This fits with other researchers' claims of beneficially enriching the gas with electrons.

To many clean-energy researchers around the world Meyer is a hero, but to jubilant skeptics he was declared a fraud after a court ruling in 1996. This writer knew Meyer, however, and knew that what Meyer told the court was true as he saw it and that he could have lost his opportunity to patent parts of his invention if he revealed its secrets to the court. The patent office was at the time investigating his technology, as were the U.S. Department of Energy and his country's military. Also, court-appointed expert witnesses who declared that his invention couldn't work may have well been experts in standard electrolysis, but not experts in what he was actually doing.

Meyer carried on his work even after the court ruling. Two years later he was at a restaurant celebrating a promise of funding when he jumped up from the table, rushed outside declaring that he had been poisoned, and died in the parking lot.

More recently, television station engineer John Kanzius learned how to use radio frequencies to release a combustible fuel from salt water. Henry Garrett, Meyer, Kanzius, and others had no qualms about violating Faraday's laws. Now they have joined Faraday by having left this physical realm and could be exchanging knowledge in an inventors' heaven. The Stan Meyer whom I knew would be telling Faraday, "You're a genius, Mike, but you didn't have the benefit of modern electronics. If you had had my voltage-intensifier circuit, society would've been running cars and generators on water long ago."

If Faraday were alive today, would he tell experimenters that their results are too good to be true and that they should give up trying? With his enthusiasm for discovery, it is likely Faraday instead would join the twenty-first-century maverick engineers who are producing hydrogen gases seven or ten times more efficiently than his nineteenth-century experiments predicted. The emerging science of super-efficient electrolysis promises a clean fuel produced at the point and time where it is needed. One of the leaders of the hydrogen-from-water experimenters' movement, Bob Boyce, says, "Water can be transformed into a perfect energy carrier. It is abundant, non-polluting and is eternal in nature. . . . After using it to gain a benefit of increased combustion efficiency and reduced emissions, we release it as clean water vapor where it will be recycled by nature. Hard to beat!"

OPEN SOURCING

Boyce generously shared the technical how-to details of his own water-as-fuel adventures and helped other hobbyists on the Internet. That type of sharing, called open sourcing, is becoming widespread. George Wiseman in western Canada was an early adopter of a "no-patent policy." Instead of patenting his fuel-efficiency discoveries, he wrote how-to books. Buyers of his books were grateful and corresponded with him about their own experiences with his inventions. The next version of each of Wiseman's books contains updated information, and the devices are improved as a result of his readers' feedback.

Boyce's story illustrates why some new-energy researchers become suspicious of the business-as-usual approach. He had worked in broadcast engineering and other electronics, so when he lived in southern Florida he opened an electronics business with a machine shop behind it. There he worked on racing boat engines and did jobs for local mini-sub researchers who were building drone boats for the government. In 1988, his business began sponsoring a small-boat race team, which led him to experimenting with running the boats on hydrogen. Boyce was separating the hydrogen from water "on the fly"—on demand in the boat as needed—but he had never heard of Puharich, Meyer, or any other of his predecessors. At first he built fairly conventional electro-lyzers. They were inefficient, but electricity consumption wasn't a big problem since races were over quickly and batteries recharged.

His anomalous breakthrough was an accident. While his two boats were in racing events, a strange coincidence caught his attention. One of his boats would leap forward with unprecedented thrust whenever its engine ran in a certain revolutions-per-minute (rpm) range. The burst of power was as if a turbocharger had kicked in. Boyce began monitoring the boat's electrical system and noticed that at those engine speeds a particular electrical waveform always appeared on the monitoring equipment. Later, he learned that the waveform was caused by electrical shorting within the boat engine's alternator.

Boat racers welcome a burst of horsepower from an engine, if they know how to control it. Boyce learned that at a specific frequency range the electrolyzer suddenly bubbled out an overpowering amount of hydrogen gas. But how could he recreate that strange and beneficial result in other electrolyzers and at all engine speeds—and reliably? This mystery intrigued Boyce so much that he became more interested in research than racing.

Eventually he learned that recreating the effect didn't require a faulty alternator. Instead, he could artificially induce the specific waveform. With signal generators, audio transformers, and amplifiers imitating that electrical signal, Boyce made test systems to reproduce the excess-hydrogen effect. The superimposed electrical frequencies seemed to cause the *resonance* effect in water molecules. Boyce began to understand why hydrogen and oxygen had separated easily whenever that waveform showed up on a scope.

RESONANCE UNLOCKS
WATER MOLECULES

Resonance is well known to musicians. When a musical instrument makes a tone at a certain frequency of vibrations, an instrument across the room picks up the vibration if one of its strings is tuned to that pitch. Another analogy for resonance is pushing a child on a swing. Small pushes, correctly timed, gradually result in impressive swinging. Similarly, sound vibrations or tiny pulses of voltage, if timed correctly, could do more work than comparatively massive amounts of electricity.

Because resonance can tear something apart, long columns of soldiers marching over a bridge break step so materials in the bridge don't resonate with their rhythmic steps and begin to fall apart. At an atomic level, when a substance is shaken rhythmically at a frequency that entrains its atomic structure, the resonating movements build up, and bonds between atoms may be more likely to break apart. Making use of resonance is one way that water-as-fuel experiments have moved beyond Faraday's electrolysis.

Another question Boyce encountered had to do with the *type* of hydrogen being released from water in his experiments. The common form of hydrogen is called diatomic—two atoms bound together. It seemed that an abundance of the form called monoatomic—single-atom hydrogen—was involved after he began making electrolytic cells with special geometries and roughing the steel plates to create more surface area.

Unknown to Boyce, there were similarities between the boxed arrays of parallel stainless steel squares he was putting together and what a scientist named William Rhodes had patented. *Also* unknown to Boyce, a Bulgarian immigrant in Australia, Yull Brown, had learned that similar apparatus did indeed release a special mix of gases.

Even without resonance effects, Boyce could double the output efficiency that Faraday had declared as the limit. What was in the special mixture? Boyce discovered that with the right frequencies, his electrolyzer could generate monoatomic as well as normal hydrogen and oxygen. When those unusual single-atom forms of hydrogen and oxygen recombined, they produced about four times the energy output of normal hydrogen and oxygen molecules. The output was stimulated into even higher efficiency by the multiple harmonic resonances that helped "tickle" water molecules apart. Boyce described the process as an electrochemical reaction, but the mixture of gases held even more mysteries than he knew at the time.

With an old Chrysler automobile hoisted on a jack stand in his shop, Boyce began testing how well a car engine ran on his hydrogen mixture. He knew how much horsepower was put out by the engine when standard processes created hydrogen gas, and now he was quadrupling its energy output. The strange gas mixture containing monoatomic hydrogen behaves like hydrogen, he learned, but it also behaves better. It apparently contains more energy.

CRIMINALS HALT PROGRESS

Boyce hadn't tried to market his breakthrough, but troubles began for him after he converted the Chrysler to run on hydrogen. He never had the chance to road test the system, because suddenly his shop became a target for criminals who broke into the building. They only vandalized or stole equipment related to the hydrogen project. The thieves took his electronics control unit and smashed the hydrolysis chamber containing the steel plates.

Boyce rebuilt a Plexiglas box and replaced the stolen electronics. However, he never finished the second unit for his engine because his shop was broken into again. Thieves stole his second prototype and an inverter he had modified. They smashed the last remaining Plexiglas box on his bench. Boyce gave up.

Walking away from the investment of time and money was wrenching. He had spent many thousands of dollars on materials and machining. Boyce later suspected that one of his hired workers might have leaked information to someone who then tried to either steal the technology or stop Boyce from working on it. He converted the boat engines back to using ordinary racing fuel and sold his race boats. In a final break-in, no damage was done inside his shop; apparently the criminals could see he was no longer experimenting with water as fuel. Boyce retired in 1991.

Eventually Boyce learned that what he had stumbled on had already been discovered and was known as Brown's gas, HHO, or hydroxyl. Ruggero Santilli has a related gaseous fuel he calls magnegas. The process of making it can result in the cleanup of sewage waters and liquid industrial wastes.

Many experimenters want to learn how to run a car solely on water, and they try to replicate Meyer's invention. Some others are using plans given out by Paul Pantone and injecting water vapor into lawn mower engines. A few mavericks are tweaking those instructions in order to run a stationary electric generator on water. The largest of such international subcultures of experimenters is the "hydroxy booster" community, perhaps because the project looks easier than a completely water-powered engine. Hydroxy gas is a nickname for oxyhydrogen, a two-to-one mix of hydrogen and oxygen gases produced from the electrolysis of water. The hydroxy experimenters' usual priority is to boost miles-per-gallon for a car, truck, or motorcycle. Improving exhaust emissions happens at the same time.

RHODES'S, BROWN'S, OR HYDROXY GAS

Rhodes was first to patent a process for making oxyhydrogen, but Brown was first to point out its unusual properties. As a result, the name Brown's gas is widely used. Progressing further in understanding the gas, Wiseman gives talks about electrically expanded water. Companies such as his Eagle Research sell equipment and the plans for making electrolyzers.

Builders of hydroxy boosters have up to now been outsiders. Their credibility was shredded when a television documentary exposed the peddling of an overhyped electrolysis system and the public received the impression that the entire field of endeavor was fraudulent. On a separate television program, an entertaining duo ridiculed the field by humorously demonstrating a poorly built electrolysis unit. The televised gadget even lacked safety features such as the flashback-arresting bubbler that responsible builders include when they build a unit. The unspoken take-away message: "Hydroxy boosters are unsafe, ineffective, and a joke."

However, neither program contacted any of the many builders of successful projects. Meanwhile the builders report hydroxy boosters producing as much as sixteen liters of nonpolluting gas per hour, using a fraction of the electricity that science books say is needed.

If the experimenters are indeed running engines on hydrogen or boosting mileage with hydrogen, something impossible is happening.

Some of the hydroxy booster outputs are said to exceed Faraday's limit by seven, ten, and even a hundred times. Who is correct?

Science writer Moray King stands in the middle of the controversy and declares that both sides are talking past each other. He says each is speaking about a different reality.

- Skeptics are correct in citing Faraday and the laws of thermodynamics.
- At the same time, hydroxy booster builders and replicators of Meyer's water fuel cell are correct in citing the excess power output that they see.
- Neither side is fully correct. Neither should be talking about power from hydrogen because, surprisingly, hydrogen is not the source of excess power in these inventions, King says.

King, author of *Tapping the Zero-Point Energy* and other books, backs up his conclusions with hundreds of references to peer-reviewed science journals. He concludes that "charged water clusters" are responsible for the anomalous results and possibly for tapping into the background energy of the universe. Meyer once told me the background energy surrounding and permeating us—whether it's called aether or zero point energy—was the ultimate source of the excess power converted by his own invention.

The hydroxy community wants academics to research their field. They welcomed Chris Eckman, an Idaho State University graduate student of nuclear engineering, when he stepped outside academia one summer weekend to attend an ExtraOrdinary Technology Conference in Utah. There he heard differing definitions of Brown's gas, so he returned to school and analyzed and experimented with the mysterious gas for two years. He shared his results with the next such conference—as a speaker at the age of twenty-four. Eckman's surprising findings included the finding that Brown's gas is not made of hydrogen—whether single-atom or two-atom molecules. Instead it is a gaseous form of water with excess electrons.

It's time to solve these water-related mysteries—for clean energy technologies. Working outside of military research facilities and corporate laboratories that don't intend to freely share their discoveries with the peoples of the world, those who are developing the most potentially world-changing energy science are often self-funded. To buy electronic parts, materials, and machining, these individuals may spend their savings and sink into credit card debt, then borrow from family and friends. Just as the personal-computer revolution began in garage workshops, much of the energy revolution is homegrown. At this time a large part of it is a grassroots movement, seeded by shared ideas and spreading over the Internet. Venture capitalists are willing to step in when an invention has gone through all the prototype stages to the completion and certified testing of a commercial product, but are not willing to fund the research-and-development stages.

ORDINARY WATER, "GREATER THAN FIRE"

There are exceptions to that widespread picture of inventors who lack funding and a team of scientists to further develop their energy breakthroughs. Dr. Randell Mills of Princeton, New Jersey, stands out as exceptional. He has a medical degree from Harvard University and studied electrical engineering at Massachusetts Institute of Technology. Mills and his colleagues have had more than eighty peer-reviewed science papers published about his new chemical process that releases energy from hydrogen atoms without combustion and without any harmful radiation.

As this is written, Mills's company, BlackLight Power Inc., has signed another commercial agreement, this time licensing an Italian energy provider to use the BlackLight process to make heat. That abundant heat can be turned into electricity. Mills has a further invention that bypasses the heat phase; the process goes directly to electricity.

BlackLight Power's tag line is "Greater than Fire." The company says its new energy source means efficient generators of electricity could be built in any size. The power density of the process competes with that of coal and nuclear fission, so the company says its energy source, when in the future placed into buyers' hands by licensees, can be used for affordable heating and electricity generation and in powering vehicles.

How did he leap from inventing for the medical field to saving the world from fuel pollution? Mathematics was the bridge; Mills developed his own unified physics theory. To test his theory's predictions, he experimented with ordinary water and a catalyst. Water is made of molecules that each bind an oxygen atom with two hydrogen atoms. His theory predicted that a hydrogen atom's electron orbit could be tightened, forming a smaller atom that he calls a "hydrino." (Other private laboratories more recently getting similar results call their process "fractional hydrogen" and seek other theories to explain it.)

Although Mills' theory is controversial, his experiments are successful and have been replicated by independent laboratories. A Rowan University team found that his heated-hydrogen-and-catalyst reaction produces far more heat than their chemists can account for with standard chemistry.

By-products of the BlackLight process are:

- Novel hydrogen compounds that are being made into useful new materials. Perhaps such discoveries will prevent the need for mining certain metals from the Earth.
- A stable hydrino gas. Highly buoyant, it floats harmlessly up through the atmosphere to space if released.

Since the process releases about two hundred times more energy than comes from *burning* an equal amount of hydrogen gas, very little water is needed to provide the hydrogen. The company says only one-millionth of a liter of water per second is needed per kilowatt of electric power. This means the system may be self-contained except for replacing the hydrogen consumed to make hydrinos.

DIRECT-TO-ELECTRICITY IS GAME-CHANGING

The company's new way to develop power by the reaction of hydrogen to form hydrinos is more efficient and cost effective. BlackLight's technology, called *Catalyst-Induced-Hydrino-Transition* (CIHT), is a unique electrochemical cell that goes directly to electricity. Capital costs are projected to be only about two percent of the cost of building conventional power systems. Fossil fuel and nuclear power plants rely on generating heat to create steam that turns turbines. The new electrochemical cell is described as suitable for generating power at any scale—small appliance to large power plant. Mills pictures a concept electric car whose drive train would be powered by an onboard process without turbines or heat engines.

Instead of the expensive standard version of a hydrogen economy—one that requires government subsidies—will we instead enjoy a low-impact fractional-hydrogen economy? Mills's invention if brought to market would make hydrogen pipelines, fueling stations, and storage tanks obsolete. BlackLight Power products and other emerging inventions also eliminate the need for costly hydrogen fuel cells.

The financial weight and political power of the dominant energy industries discourage some energy innovators but haven't stopped Mills. Inventors hear stories about opposition to some energy innovators' efforts. Mills however does not report having suffered the physical harassment that some inventors have endured. Opposition came instead in the form of public ridicule from a sector that relies on government grants. For instance, one physicist connected to the hot-fusion research establishment declared that science already knows everything about the hydrogen atom. Mills didn't waste his time arguing with such critics. More seriously, however, at a point when patenting was crucial to getting investors there was evidence of behind-the-scenes interference with Mills's patent application.

The inventive doctor found ways to get money for his energy research. Since Mills's mathematics solves equations for figuring out the physical structure of electrons in atoms and molecules, a subsidiary of BlackLight is selling a separate product that does not threaten the energy establishment. Biological and industrial scientists buy the company's software product to map properties of various metals and chemicals.

Meanwhile, BlackLight Power plans to license its heat-producing energy process. Licensees might retrofit existing power plants or hire architects, engineering firms, and equipment manufacturers to build large or small power plants.

WHERE CAN THE HUMAN FAMILY GO WITH THIS?

There are more breakthrough water-and-energy technologies than one chapter can cover. For instance, cleaning up polluted rivers can be done by a ready-to-use Vapor Condensation Distillation unit developed by a California man, Stephan Sears. His company's plans include solving water problems in undeveloped countries where groundwater, rivers, lakes, or wells have been polluted. Technically it would be easy—combining the unit with a small photovoltaic (solar cell) panel, small windmill, or a run-of-the-river turbine that generates electricity. Only a small fraction of the electricity is needed to run the water distillation process; the unit uses very little power because it recycles its heat and has other exceptionally efficient features.

Many of the researchers look for grassroots support as the way to get their innovations into the marketplace. The people of Earth are a superpower themselves, if united.

The alternative to worldwide cooperation is grim—segments of the human family each allowing their national leaders to use the fact of water shortages to justify continued warring over scarce resources. The public can insist on these outside-the-box breakthroughs. Currently government agencies buy only politically safe research—lesser advances that don't really threaten the oil industry. The public could be asking, "Why do we allow technicians to dig and drill into our planet for black fuels? Don't they know what can be done with just a bucket of precious water? They could be powering or heating whatever is needed, without burning anything."

THE GATEWAY
TO INFINITE ENERGY

A new scientific truth does not triumph by convincing its opponents and making them see the light, but rather because its opponents die, and a new generation grows up that is familiar with it.

MAX PLANCK

Electric power is everywhere present in unlimited quantities and can drive the world's machinery without the need of coal, oil, gas, or any other of the common fuels.

NIKOLA TESLA

Following the decimation of Tesla around the turn of the century, similar tactics have continued against follow-on inventors who discovered overunity systems and attempted to complete them and bring them to market. The suppression continues to this day, as can be attested by several living overunity inventors and inventor groups. For more than a century there has indeed been a giant, unwritten conspiracy of some of the most powerful cartels on Earth, to continue the curtailment of the electrical engineering model and practice, and to continue to suppress overunity inventions and inventors.

More than a century ago and along with its very birthing, our "modern" classical electrodynamics and electrical engineering science was deliberately mutilated and crippled, specifically so that COP > 1.0 and self-powering electrical systems—asymmetrically powering loads extracted from "free EM wind energy flows" from the vacuum/space itself—would never be known or developed by our electrical engineers. . . . It has directly prevented struggling nations having no oil or gas resources from achieving a modern economy (which is based on cheap energy). This has left those nations impoverished, with their peoples starving and miserable and disease-wracked. Hundreds of millions of deaths from starvation and disease have resulted worldwide. It has "welded into our minds and our very brains" the mistaken notion that—other than a wee bit of wind power, water power, and solar power—we can only have "energy from consumption of fuel."

Our electrical engineers and scientists today are totally unaware that every generator already pours out more than a trillion times as much EM energy flow from the vacuum, as is in the mechanical energy flow we input to the generator shaft.

THOMAS E. BEARDEN

The release of the energy of the atom is as yet in an extremely embryonic stage; humanity little knows the extent or the nature of the energies which have been tapped and released. There are many types of atoms, constituting the "world substance"; each can release its own type of force; this is one of the secrets which the new age will in time reveal. . . . I would call your attention to the words, "the liberation of energy." It is liberation which is the keynote of the new era, just as it has ever been the keynote of the spiritually oriented aspirant. . . . This liberating energy will usher in the new civilization, the new and better world and the finer, more spiritual conditions. The highest dreams of those who love their fellowmen can become practical possibilities through the right use of this liberated energy, if the real values are taught, emphasized and applied to daily living.

DJWHAL KHUL

8

IMAGINE A FREE-ENERGY FUTURE FOR ALL OF HUMANITY

Steven M. Greer, M.D.

At no time, when the astronauts were in space were they alone: there was a constant surveillance by UFOs.

SCOTT CARPENTER, U.S. ASTRONAUT
THE FOURTH MAN IN SPACE

In the councils of Government, we must guard against the acquisition of unwarranted influence, whether sought or unsought, by the Military Industrial Complex. The potential for the disastrous rise of misplaced power exists, and will persist. We must never let the weight of this combination endanger our liberties or democratic processes. We should take nothing for granted. Only an alert and knowledgeable citizenry can compel the proper meshing of the huge industrial and military machinery of defense with our peaceful methods and goals so that security and liberty may prosper together.

PRESIDENT DWIGHT D. EISENHOWER, 1961

■■■

Technological progress in the areas of advanced physics and electromagnetic systems, if appropriately supported, will enable humanity to live on the Earth with a minimal footprint with genuine long-term sustainability. For over one hundred years, these advanced concepts in energy generation have either been ignored or actively suppressed due to the power of fossil-fuel-based economic and industrial interests.

Imagine a world where every home and village has its own clean source of electrical energy, free from the cost of fossil fuels, nuclear power, or a centralized electric grid. Imagine every means of transportation running off of clean power plants, using no source of fuel and creating no pollution.

Imagine the developing world blossoming with these new technologies and the equatorial rain forests protected from slash-and-burn subsistence farming and logging.

Imagine all intercity transportation moving above the ground and the millions of acres now paved over with highways freed for productive agriculture and recreation.

Imagine all manufacturing being clean-fuel sourced, using no-cost or low-cost energy.

Imagine the possibility of 100 percent recycling because the energy cost of transporting recycled materials, processing them, and scrubbing pollution out of the air and water approaches zero.

Make no mistake, the changes would be immense. Yet we have these technologies today, hidden away in ultrasecret military-industrial programs, paid for by the taxpayers, yet of which the public knows nothing. What ethic can justify such secrecy in today's world?

To understand where we are going we need to take a look at the history of "free energy."

OUR COSMIC NEIGHBORHOOD

For most people, the question of whether or not we are alone in the universe is a mere philosophical musing—something of academic interest but of no practical importance. Even evidence that we are currently being visited by nonhuman advanced life-forms seems to many to be an irrelevancy in a world of global warming, crushing poverty, and the threat of war. In the face of real challenges to the long-term human future, the question of UFOs, extraterrestrials, and secret government projects is a mere sideshow, right? Wrong! Catastrophically wrong.

The evidence and testimony presented in the following pages establishes the following:

- That we are indeed being visited by advanced extraterrestrial civilizations and have been for some time
- That this is the most classified, compartmentalized program within the United States and many other countries
- That these projects have, as warned in 1961 by President Eisenhower, escaped legal oversight and control in the United States, the United Kingdom, and elsewhere
- That advanced spacecraft of extraterrestrial origin, called extraterrestrial vehicles (ETVs) by some intelligence agencies, have been downed, retrieved, and studied since at least the 1940s and possibly as early as the 1930s
- That significant technological breakthroughs in energy generation and propulsion have resulted from the study of these objects (and from related human innovations dating as far back as the time of Nicola Tesla) and that these technologies use a new physics not requiring the burning of fossil fuels or ionizing radiation to generate unlimited amounts of energy
- That classified, above-top-secret projects possess fully operational antigravity propulsion devices and new energy–generation systems that, if declassified and put into peaceful uses, would empower a new human civilization without want, poverty, or environmental damage.

Those who doubt these assertions should carefully read my book, *Disclosure*,[1] which includes the testimony of dozens of military and government witnesses whose statements clearly establish these facts. Given the vast and profound implications of these statements, whether one accepts or doubts them, all must demand that Congressional hearings be convened to get to the truth of this matter. For nothing less than the human future hangs in the balance.

IMPLICATIONS
FOR THE ENVIRONMENT

I have identified insiders and scientists who can prove, in open Congressional hearings, that we do in fact possess classified energy-generation and antigravity propulsion systems capable of completely and permanently replacing all forms of currently used energy-generation and transportation systems. These devices access the ambient electromagnetic and so-called zero point energy states to produce vast amounts of energy without any pollution. Such systems essentially generate energy by tapping into the ever-present quantum vacuum energy state—the baseline energy from which all energy and matter is fluxing. All matter and energy is supported by this baseline energy state, and it can be tapped through unique electromagnetic circuits and configurations to generate huge amounts of energy from space-time all around us. These are *not* perpetual motion machines, nor do they violate the laws of thermodynamics—they merely tap an ambient energy field all around us to generate energy.

This means that such systems do not require fuel to burn or atoms to split or fuse. They do not require central power plants, transmission lines, and the related multi-trillion-dollar infrastructure required to electrify and power remote areas of India, China, Africa, and Latin America. These systems are site specific: they can be set up at any location and generate energy. Essentially this constitutes the definitive solution to the vast majority of environmental problems facing our world.

The environmental benefits of such a discovery can hardly be overstated, but a brief list includes:

- The elimination of oil, coal, and gas as sources of energy generation, thus the elimination of air and water pollution related to the transport and use of these fuels. Oil spills, global warming, illnesses from air pollution, acid rain, and other ramifications can and must be ended within ten to twenty years.

- Resource depletion and geopolitical tensions arising from competition for fossil fuel resources will end.

- Traditional technologies exist today to scrub manufacturing effluent to zero or near-zero emissions for both air and water, but they use a great deal of energy and thus are considered too costly to fully use. Moreover, since they are energy intensive, and our energy systems today create most of the air pollution in the world, a point of diminishing return for the environment is reached quickly. That equation is dramatically changed when industries are able to tap vast amounts of free energy (there is no fuel to pay for—only the device, which is no more costly than other generators), and those systems create no pollution.

- Energy-intensive recycling efforts will be able to reach full application since the energy needed to process solid waste will again be free and abundant.

- Agriculture, which is currently very energy dependent and polluting, can be transformed to use clean, nonpolluting sources of energy.

- Desertification can be reversed and world agriculture empowered by using desalinization plants, which are currently very energy intensive and expensive but will become cost efficient once able to use these new, nonpolluting energy systems.

- Air travel, trucking, and intercity transportation systems will be replaced with new energy and propulsion technologies (antigravity systems allow for silent above-surface movement). No pollution will be generated, and costs will decrease substantially since the energy expenses will be negligible. Additionally, mass transportation in urban areas can use these systems to provide silent, efficient intracity movement.

- Noise pollution from jets, trucks, and other modes of transportation will be eliminated by the use of these silent devices.

- Public utilities will not be needed since each home, office, and factory will have a self-contained device able to generate whatever energy is needed. This means unsightly transmission lines that are subject to storm damage and power interruption will be a thing of the past. Underground gas pipelines, which frequently rupture or leak and damage Earth and water resources, will not be needed at all.

- Nuclear power plants will be decommissioned, and the technologies needed to clean up such sites will be available. Classified technologies do exist to neutralize nuclear waste.

Utopia? No, because human society will always be imperfect but perhaps not as dysfunctional as it is today. These technologies are real. Antigravity is a reality, and so is free-energy generation. This is not a fantasy or a hoax. Do not believe those who say that this is not possible: they are the intellectual descendants of those who said the Wright brothers would never fly.

Current human civilization has reached the point of being able to commit "planeticide": the killing of an entire world. We can and we must do better. These technologies exist, and every single person who is concerned about the environment and the human future should call for urgent hearings to allow these technologies to be disclosed and declassified and safely applied.

IMPLICATIONS FOR SOCIETY AND WORLD POVERTY

From the above, it is obvious that these technologies that are currently classified would enable human civilization to achieve true sustainability. Of course, in the near term, we are talking about the greatest social, economic, and technological revolution in human history—bar none. I will not minimize the world-encompassing changes that would inevitably attend such disclosures. Having dealt with this issue for much of my adult life, I am acutely aware of how immense these changes will be.

Aside from the singular realization that *Homo sapiens* are not the only—or most advanced—creatures in the universe, the release of these secret technologies will present humanity with the most profound choices and opportunities in known history. If we do nothing, our civilization will collapse environmentally, economically, geopolitically, and socially. In ten to twenty years, fossil fuel and oil demand will outstrip supply significantly—and then it is the *Mad Max* scenario where everyone is warring over the last barrel of oil. It is likely that this geopolitical and social collapse will precede any environmental catastrophe.

The disclosure of these new technologies will give us a new, sustainable civilization. World poverty will be eliminated within our lifetimes. With the advent of these new energy and propulsion systems, no place on Earth will need to suffer from want. Even the deserts will bloom . . .

Once abundant and nearly free energy is available in impoverished areas for agriculture, transportation, construction, manufacturing, and electrification, there is no limit to what humanity can achieve. It is ridiculous—obscene even—that mind-boggling poverty and famine exist in the world while we sit on classified technologies that could completely reverse this situation. So why not release these technologies now? The social, economic, and geopolitical order of the world would be greatly altered. Every deep insider with whom I have met has emphasized that this would be the greatest change in known human history. The matter is so highly classified not because it is so silly, but because its implications are so profound and far reaching. By nature, those who control such projects do not like change. And here we are talking about the biggest economic, technological, social, and geopolitical change in known human history. Hence, the status quo is maintained, even as our civilization hurtles toward oblivion.

But by this argument, we would have never had the Industrial Revolution, and the Luddites would have reigned supreme to this day.

An international effort to minimize disruption to the economy and to ease the transition to this new social and economic reality is needed. We can do this, and we must do this. Special interests in certain oil, energy, and economic sectors need to be reined in and at the same time treated compassionately: nobody likes to see their power and empire crumble. Nations very dependent on the sale of oil and gas will need help diversifying, stabilizing, and transitioning to a new economic order.

The United States, Europe, China, and Japan will need to adjust to a new geopolitical reality as well: as currently poor but populous countries dramatically develop technologically and economically, they will demand—and will get—a meaningful seat at the international table. And this is as it should be. But the international community will need to put in place safeguards to prevent such potential geopolitical rapprochement between the first and third worlds from devolving into bellicose and disruptive behavior on the part of nations, now or in the future.

The United States in particular will need to lead through strength but avoid the current trend toward domination. Leadership and domination are not the same, and the sooner we learn the difference the better off the world will be. There can be international leadership without domination and hegemony, and the United States needs to realize these distinctions if it is to provide much-needed leadership on this issue.

These technologies, because they will literally and figuratively decentralize power, will enable the billions living in misery and poverty to enter a world of new abundance. And with economic and technological development, education will rise and birthrates will fall. It is well known that as societies become more educated, prosperous, and technologically advanced—and women take an increasingly equal role in society—the birthrate falls and population stabilizes. This is a good thing for world civilization and the future of humanity.

With each village cleanly electrified, agriculture empowered with clean and free energy, and transportation costs lowered, poverty will dramatically fall in the world. If we act now, we could effectively eliminate all poverty in the world as we know it today. We only need the courage to accept these changes and the wisdom to steer humanity safely and peacefully into a new time.

IMPLICATIONS FOR
WORLD PEACE AND SECURITY

A good many years ago, I was discussing this subject with the former chairman of the Senate Committee on Foreign Relations, the late Senator Claiborne Pell. He explained to me that he had been in Congress since the 1950s but had never been briefed on this subject. I told him that the nature of these black projects has resulted in most of our leaders being left out of any decision making on this subject, and what a shame this is. I said, "Senator Pell, all that time you were chairman of Foreign Relations, you were deprived of the opportunity to deal with the ultimate foreign relations challenge," and I pointed to the stars above our heads. He said, "You know, Dr. Greer, I am afraid that you are right."

It is true that our great diplomats and wise elders, such as Senator Pell, President Jimmy Carter, and other international leaders, have been specifically and deliberately prevented from having access to or control over this subject. This is a direct threat to world peace. In the vacuum of secrecy, operatives supervised by neither the people, the people's representatives, the United Nations (UN), nor any other legitimate entity have taken actions that directly threaten world peace.

For instance, testimony corroborated by multiple military witnesses who did not know one another and who had no opportunity for collusion has shown that the United States and other countries have engaged ETVs in armed attack, in some cases leading to the downing of these vehicles. As I said to Mrs. Boutros Ghali, wife of the then-secretary general of the UN, if there is even a 10 percent chance that this is true, then this constitutes the gravest threat to world peace in human history.

Having personally interviewed numerous credible military and aerospace officials with direct knowledge of such actions, I am certain that we have done this. Why? Because these unknown vehicles have traversed our airspace without our permission and because we wanted to acquire their technology. Nobody has asserted that there is an actual threat to humanity from these objects: it seems to me that any civilization capable of routine interstellar travel could terminate our civilization in a nanosecond if that was their intent. That we are still breathing the free air of Earth is abundant testimony to the nonhostile nature of these ET civilizations.

We have also been informed that the so-called Star Wars (or National Missile Defense System) effort has really been a cover for black project deployment of weapon systems to track, target, and destroy ETVs as they approach Earth or enter Earth's atmosphere. No less a figure than Werner Von Braun warned on his deathbed of both the reality and the madness of such a scheme, apparently to no avail, according to Carol Rosin, his former spokesperson.

Well, unless we change directions we are likely to end up where we are going.

With the types of weapons currently in the covert arsenal—weapons more fearsome even than thermonuclear devices—there is no possibility of a survivable conflict. Yet in the darkness of secrecy, actions have been taken on behalf of every human that may endanger our future. Only a full, honest disclosure will correct this situation. It is not possible for me to convey in words the urgency of this disclosure.

For ten years I worked as an emergency room doctor and saw how every possible object can be used as a weapon. Every technology, unless guided by wisdom and a desire for that good and peaceful future—the only future possible—will be used for conflict. Those responsible for super-secret projects who answer to no legally constituted body—not the UN, not the U.S. Congress, not the British Parliament—must not be allowed to continue to act in this way on behalf of humanity.

One of the greatest dangers of extreme secrecy is that it creates a hermetically sealed, closed system impervious to the free and open exchange of ideas. In such an environment, it is easy to see how grave mistakes can be made. For instance, the testimony I've gathered shows that these ETVs became very prominent after we developed our first nuclear weapon and began to go into space. There were multiple events—corroborated by numerous credible military officials—of these objects hovering over and even neutralizing intercontinental ballistic missile sites.

A closed military view of this might be to take offense, engage in countermeasures, and attempt to down such objects. In fact, this would be the normal response. But what if these ET civilizations were simply saying, "Please do not destroy your beautiful world—and know this: we will not let you go into space with such madness and threaten others?" An event showing concern and even a larger cosmic wisdom could be construed over and over again as an act of aggression. Such misunderstandings and myopia are the stuff wars are made of.

Whatever our perceptions of these visitors, there is no chance that misunderstandings can be resolved through violent engagement. To contemplate such madness is to contemplate the termination of human civilization.

It is time for our wise elders and our levelheaded diplomats—men and women of the caliber of Senator Pell—to be put in charge of these weighty matters. To leave this in the hands of a clique of unelected, self-appointed, and unaccountable officials in covert operations is the greatest threat to U.S national security and world security in history. Eisenhower was right, but nobody was listening.

In light of testimony showing that covert actions have been taken that involved violent engagement of these visitors, it is imperative that the international community in general and the U.S. Congress and president in particular do the following:

- Convene hearings to assess the risks to national and international security posed by the current covert management of the subject
- Enforce an immediate ban on weapons in space and specifically ban the targeting of any extraterrestrial objects since such actions are unwarranted and could endanger the whole of humanity
- Develop a special diplomatic unit to interface with these extraterrestrial civilizations and to foster communication and peaceful relations
- Develop a suitably empowered and open international oversight group to manage human-extraterrestrial relations and ensure peaceful and mutually beneficial interactions
- Support international institutions that can ensure the peaceful use of those new technologies related to advanced energy and propulsion systems.

In addition to the above, a less obvious—but perhaps equally pressing—threat to world peace arises from the fact that the covert control of this subject has resulted in the world being deprived of the new energy and propulsion technologies discussed earlier.

World poverty and a widening gap between rich and poor are serious threats to world peace and would be corrected by the disclosure and peaceful application of these technologies. The real threat of war over a shrinking supply of fossil fuels in the next ten to twenty years further underscores the need for this disclosure. What happens when the 4 billion people living in poverty want cars, electricity, and other modern conveniences—all of which depend on fossil fuels? To any thinking person, it is obvious we must transition quickly to the use of these now-classified technologies; they are powerful solutions already sitting on a shelf.

Of course, a number of insiders have pointed out that these technologies are not your grandfather's Oldsmobile: they are technological advances that, like any other, could be put to violent uses by terrorists, bellicose nations, and madmen. Here we enter a catch-22: if these technologies are not forthcoming soon, we will face a certain meltdown in human civilization and the environment, but if disclosed, immensely powerful new technologies will be out there for possible destructive uses.

In the near term, it is prudent to view humanity as likely to use any new technology violently. This means that international agencies must be created to ensure—and enforce—the exclusively peaceful use of such devices. The technologies exist today to link every such device to a Global Positioning System (GPS) monitor that could be used to disable or render useless any device tampered with or used for anything but peaceful power generation and propulsion. These technologies should be regulated and monitored. And the international community must mature to a level of competence to ensure their exclusively peaceful use. Any other use should be met with overwhelming resistance by every other nation on Earth.

Such a pact is the necessary next step. Maybe someday, humanity will live in peace without the need for such controls, but for now the situation is like that of chained dogs: some strong leashes are warranted and I believe are essential.

It is also essential that such concerns not be a rationale for further delaying the disclosure of these technologies. We have the knowledge and means to ensure their safe and peaceful use, and these must be applied soon if we are to avoid further degradation of the environment and an escalation of world poverty and conflict.

In the final analysis then, we are faced with a social and spiritual crisis that transcends any technological or scientific challenge. The technological solutions exist, but do we possess the will, wisdom, and courage to apply them for the common good? The more one contemplates this matter, the more it is obvious we have but one possible future: peace for man, peace on Earth, and peace in space—a universal peace, wisely enforced. For every other path truly leads to ruin.

This then is the greatest challenge of the current era. Can our spiritual and social resources rise to this challenge? Nothing less than the destiny of the human race hangs in the balance.

9

ENERGY TECHNOLOGIES FOR THE TWENTY-FIRST CENTURY

Theodore C. Loder III, Ph.D.

NOTE: This essay was originally presented at the fortieth American Institute of Aeronautics and Astronautics Inc. (AIAA) Aerospace Science Meeting and Exhibit, Reno Nev., January 14–16, 2002, Paper no. AIAA-1131.

> *Any sufficiently advanced technology is indistinguishable from magic.*
>
> ARTHUR C. CLARKE

Following is a review of the development of antigravity research in the United States that notes how research activity seemed to disappear by the mid-1950s. It then addresses recently reported scientific findings and witness testimonies that show us this research and technology are alive and well and very advanced. The revelations of findings in this area will dramatically alter our twentieth-century view of physics and technology and must be considered in planning for both energy and transportation needs in the twenty-first century.

223

HISTORICAL BACKGROUND

Townsend Brown's Technology of Electrogravitics[1]

In the mid-1920s Townsend Brown (see chapter 5, "T. Townsend Brown" by Jeane Manning) discovered that electric charge and gravitational mass are coupled.[2] He found that when a capacitor is charged to a high voltage, it has a tendency to move toward its positive pole. His findings, which became known as the Biefeld-Brown effect, were opposed by conventional-minded physicists of his time.

The Pearl Harbor Demonstration

Around 1953, Brown conducted a demonstration for military top brass. He flew a pair of three-foot-diameter discs around a fifty-foot course tethered to a central pole. Energized with 150,000 volts and emitting ions from their leading edge, they attained speeds of several hundred miles per hour. The subject was thereafter classified.

Project Winterhaven

Brown submitted a proposal to the Pentagon for the development of a Mach 3 disc-shaped electrogravitic fighter craft. Drawings of its basic design are shown in one of his patents. They are essentially large-scale versions of his tethered test discs.

REVIEW OF ISSUES
FROM THE 1950S

Once Brown's findings became well known, some scientists began to openly speak about the flying technology of UFOs, which had been observed extensively since the 1940s. None other than Professor Hermann Oberth, considered by some to be one of the fathers of the space age, who later worked in the United States with renowned rocket scientist Werner von Braun, the Army Ballistic Missile Agency, and NASA, stated the following in 1954: "It is my thesis that flying saucers are real and that they are space ships from another solar system." Perhaps of more interest to our present discussion on propulsion, he then stated, "They are flying by the means of artificial fields of gravity.

... They produce high-tension electric charges in order to push the air out of their paths, so it does not start glowing, and strong magnetic fields to influence the ionized air at higher altitudes. First, this would explain their luminosity. ... Secondly, it would explain the noiselessness of UFO flight. ...[3]

We now know he was fundamentally correct in his assessment.

In 1956, a British research company, Aviation Studies (International) Ltd., published a classified report on electrogravitics systems examining various aspects of gravity control. They summarized the pioneering work of Townsend Brown and then described the use of electrogravitic thrust as follows: "The essence of electrogravitics thrust is the use of a very strong positive charge on one side of the vehicle and a negative on the other. The core of the motor is a condenser and the ability of the condenser to hold its charge (the K-number) is the yardstick of performance. With air as 1, current dielectrical materials can yield 6 and use of barium aluminate can raise this considerably, barium titanium oxide (a baked ceramic) can offer 6,000 and there is a promise of 30,000, which would be sufficient for supersonic speed."[4]

In one of their conclusions, based on Brown's work, they suggested, "Electrostatic energy sufficient to produce a Mach 3 fighter is possible with megavolt energies and a k of over 10,000."[5]

In spite of Brown's solid research, they later stated, "One of the difficulties in 1954 and 1955 was to get aviation to take electrogravitics seriously. The name alone was enough to put people off."[6] It seems that is as true today as it was in the 1950s.

A report by another British company, Gravity Rand Ltd., in 1956, agrees with this assessment and states, "To assert electrogravitics is nonsense is as unreal as to say it is practically extant. Management should be careful of men in their employ with a closed mind or even partially closed mind on the subject."[7]

However, a trade press magazine, *The Aviation Report,* made numerous references to antigravity projects and listed many of the companies pursuing research in this area. Quotes from *The Aviation Report* listed in the Aviation Studies (International) Ltd. report[8] are suggestive of what was going on behind the scenes.

In 1954 they predicted, "Progress has been slow. But indications are now that the Pentagon is ready to sponsor a range of devices to help further knowledge. . . . Tentative targets now being set anticipate that the first disk should be complete before 1960 and it would take the whole of the 'sixties to develop it properly, even though some combat things might be available ten years from now."[9]

During this time period many of the major defense and technology companies were cited as either having research projects or activities in this new field. For example, "Companies studying the implications of gravitics are said, in a new statement, to include Glenn Martin, Convair, Sperry-Rand, and Sikorsky, Bell, Lear Inc. and Clark Electronics. Other companies who have previously evinced interest include Lockheed, Douglas and Hiller."[10]

Other of these reports mentions AT&T, General Electric, as well as Curtiss-Wright, Boeing, and North American as having groups studying electrogravitics.

During the same time period, the Gravity Rand report notes, "Already companies are specializing in evolution of particular components of an electogravitics disk."[11]

However, in the area of predictions, *The Aviation Report* stated the following based on an extrapolation of technology development: "Thus this century will be divided into two parts—almost to the day. The first half belonged to the Wright Brothers who foresaw nearly all the basic issues in which gravity was the bitter foe. In part of the second half, gravity will be the great provider. Electrical energy, rather irrelevant for propulsion in the first half becomes a kind of catalyst to motion in the second half of the century."[12]

Looking back it is easy to say they missed the mark. Did they really miss it by a half a century? Reading through these reports it is quite obvious that there was much interest in antigravity among a number of very high-profile companies, as well as in the Department of Defense. What happened to this interest, and why was it all downplayed during the following four-plus decades? After all, Brown had shown there is a demonstrable connection between high-voltage fields and gravity. Why has it taken until the 1990s for more than just a few scientists to look at these results and publish on them in the open literature? A review of recent statements by former military personnel and civilians connected to covert projects begins to shed light on research activity in these areas over the last half-century. And it appears there had been significant breakthroughs during this time period, well shielded from both the scientific and public eye.

RECENT SCIENTIFIC DEVELOPMENTS

In this section we consider developments in the antigravity field since the late 1980s and why the confluence of scientific findings and the testimony of witnesses associated with the military and covert groups indicate that a gravity solution with technological implications has been found.

Although general relativity has not been able to explain Brown's electrogravitic observations, or any other antigravity phenomenon, the recent physics methodology of quantum electrodynamics (QED) appears to offer the theoretical framework to explain electrogravitic coupling. Recent papers by members of the Institute for Advanced Study Alpha Foundation are putting a solid theoretical foundation onto the antigravity effects within the theory of electrodynamics and include papers by Myron W. Evans[13] and P. K. Anastasovski et al.[14]

Earlier, in a 1994 breakthrough paper, Alcubierre showed that superluminal space travel is, in principle, physically possible and will not violate the tenets of the theory of relativity.[15] Harold Puthoff later analyzed these findings in light of the present SETI (Search for Extraterrestrial Intelligence) paradigms that insist that we could not be visited by extraterrestrial civilizations because of the speed-of-light limitations dictated by the general relativity theory.[16] He suggests that superluminal travel is indeed possible. This leads to reduced-time interstellar travel and the possibility of extraterrestrial visitation, which our limited understanding of physics and scientific arrogance has "forbidden" in some sectors for most of the twentieth century.

The second aspect of these physics findings deals with the zero point or vacuum-state energy shown by the Casimir effect,[17] which predicts that two metal plates close together attract each other due to imbalance in the quantum fluctuations. The implications of this zero point or vacuum-state energy are tremendous and are described in several papers by Puthoff,[18] starting during the late 1980s. Bearden and colleagues have also written extensively on the theoretical physics of zero point energy[19] and additionally have described various technological means of extracting this energy (for example see the recent paper by Anastasosvki et al.).[20] A theoretical book on zero point energy (and antigravity) was published by Bearden in 2002.[21] There is significant evidence that scientists since Tesla have known about this energy, but that its existence and potential use have been discouraged and indeed suppressed over the past half-century or more.[22]

The coupling of the electrogravitic phenomena observations and the zero point energy findings is leading to a new understanding of both the nature of matter and of gravity. This is just now being discussed in scientific journals (though some evidence suggests it has been understood for decades within the black project covert community). The question that is being addressed is, What keeps the universe running? Or more specifically, where do electrons get their energy to keep spinning around atoms? As electrons change state they absorb or release energy, and where does it come from? The simplistic answer is that it is coming from the vacuum state. Puthoff describes the process as follows: "I discovered that you can consider the electron as continually radiating away its energy as predicted by classical theory, but *simultaneously absorbing a compensating amount* of energy from the ever-present sea of zero point energy in which the atom is immersed. An equilibrium between these two processes leads to the correct values for the parameters that define the lowest energy, or ground-state orbit (see "Why atoms don't collapse," *New Scientist*, July 1987). Thus there is a *dynamic equilibrium* in which the zero-point energy stabilizes the electron in a set ground-state orbit. It seems that the very stability of matter itself appears to depend on an underlying sea of electromagnetic zero-point energy."[23]

Furthermore, it appears that it is the spinning of electrons that provides inertia and mass to atoms. These theories, linking electron spin, zero point energy, mass, and inertia, have been presented in a number of recent papers, such as those by Haisch and colleagues, and provide us with a possible explanation of the Biefeld-Brown effect.[24] It appears that an intense voltage field creates an electromagnetic barrier that blocks the atomic structure of an atom from interacting with the zero point field. This slows down the electrons, reducing their gyroscopic effect, and thus reducing atomic mass and inertia, making them easier to move around.

EVIDENCE OF EXTENSIVE ANTIGRAVITY TECHNOLOGY

The B-2 Advanced Technology Bomber

In 1993, Paul LaViolette wrote a paper[25] discussing the B-2 bomber and speculating on its probable antigravity propulsion system, based on a solid understanding of electrogravitics,[26] the aircraft's design, and the materials used in its manufacture. It appears the craft is using a sophisticated form of the antigravity principles first described by Brown. Support for this thesis came from *Aviation Week and Space Technology* (March 9, 1992), which reported that the B-2 bomber electrostatically charges its leading edge and its exhaust stream. Their information had come from a small group of former black project research scientists and engineers, suggesting the B-2 uses antigravity technology. This information was supported by Bob Oechsler, an ex-NASA mission specialist who had publicly made a similar claim in 1990. These findings support the contention that there have been major developments in the area of antigravity propulsion that are presently being applied in advanced aircraft.

LaViolette later states the obvious—that "the commercial airline industry could dramatically benefit with this technology which would not only substantially increase the miles per gallon fuel efficiency of jet airliners, but would also permit high-speed flight that would dramatically cut flight time."[27]

THE DISCLOSURE PROJECT
WITNESSES

On May 9, 2001, a private organization, the Disclosure Project,[28] held a press conference at the National Press Club in Washington, D.C. (see chapter 8, "Imagine a Free Energy Future for All of Humanity" by Steven M. Greer, M.D.). They presented nearly two dozen witnesses including retired army, navy, and air force personnel, a top FAA official, members of various intelligence organizations including the CIA and NRO, and industry personnel, all of whom who had witnessed UFO events or had inside knowledge of government or industrial activities in this area. They also produced a briefing document[29] for members of the press and Congress and a book[30] that includes the testimony of nearly seventy such witnesses from a pool of hundreds. Although they all spoke of the reality of the UFO phenomena, many also spoke of covert projects dealing with antigravity, zero point energy technologies, and development of alien reproduction vehicles (ARVs) by U.S. black project and covert interests. The following excerpted quotes from these witnesses support the above contentions.

Dan Morris is a retired air force career master sergeant who was involved in the extraterrestrial projects for many years. After leaving the air force, he was recruited into the supersecret National Reconnaissance Organization (NRO), during which time he worked specifically on extraterrestrial-connected operations. He had a cosmic top-secret clearance (thirty-eight levels above top secret), which, he states, no U.S. president, to his knowledge, has ever held.

"UFOs are both extraterrestrial and manmade. Well, the guys that were doing the UFOs, they weren't sleeping, and Townsend Brown was one of our guys who was almost up with the Germans. So we had a problem. We had to keep Townsend Brown—what he was doing on anti-gravity electromagnetic propulsion secret." He then describes a type of zero point energy device.

"Well, if you have one of these units that's about sixteen inches long and about eight inches high and about ten inches wide, then you don't need to plug into the local electric company. These devices burn nothing. No pollution. It never wears out, because there are no moving parts.

What moves are electrons, in the gravity field, in the electronic field, and they turn in opposite directions, okay?"[31]

"Dr. B." (name withheld since he still works in this area) is a scientist and engineer who has worked on top-secret projects almost all his life. Over the years he has directly worked on or had involvement with such projects involving antigravity, chemical warfare, secure telemetry and communications, extremely high-energy space-based laser systems, and electromagnetic pulse technology.

"Anti-gravity. As a matter of fact, I used to go out to the Hughes in Malibu. They had a big think tank up there. Big anti-gravity projects; I used to talk to them out there. I'd give them ideas, because they bought all my equipment. But the American public will never, never hear about that. . . . This flying disc has a little plutonium reactor in it, which creates electricity, which drives these anti-gravity plates. We also have the next level of propulsion, it is called virtual field, which are called hydrodynamic waves."[32]

Captain Bill Uhouse served ten years in the Marine Corps as a fighter pilot and four years with the air force at Wright-Patterson AFB as a civilian doing flight testing of exotic experimental aircraft. Later, for the next thirty years, he worked for defense contractors as an engineer of antigravity propulsion systems, on flight simulators for exotic aircraft—and on actual flying discs.

"I don't think any flying disc simulators went into operation until the early 1960s—around 1962 or 1963. The reason why I am saying this is because the simulator wasn't actually functional until around 1958. The simulator that they used was for the extraterrestrial craft they had, which is a 30-meter one that crashed in Kingman, Arizona, back in 1953 or 1952.

"We operated it with six large capacitors that were charged with a million volts each, so there were six million volts in those capacitors. . . . There weren't any windows. The only way we had any visibility at all was done with cameras or video-type devices. . . . Over the last 40 years or so, not counting the simulators—I'm talking about actual craft—there are probably two or three-dozen, and various sizes that we built."[33]

"A. H.," formerly of Boeing Aerospace, is a person who has gained significant information from inside the UFO extraterrestrial groups within our government, military, and civilian companies. He has friends at the NSA, CIA, NASA, JPL, ONI, NRO, Area 51, the air force, Northrup, Boeing, and others.

"Most of the craft operate on antigravity and electrogravitic propulsion. We are just about at the conclusion state right now regarding antigravity. I would give it maybe about 15 years and we will have cars that will levitate using this type of technology. We're doing it up at Area 51 right now. That's some of the stuff that my buddy worked on up at Area 51 with Northrup, who lives now in Pahrump, Nevada. We're flying antigravity vehicles up there and in Utah right now."[34]

Lieutenant Colonel John Williams entered the air force in 1964 and became a rescue helicopter pilot in Vietnam. He has an electrical engineering degree and was in charge of all the construction projects for the Military Air Command. During his time in the military he knew that there was a facility inside of Norton Air Force Base in California that no one was to know about.

"There was one facility at Norton Air Force Base that was close hold—not even the wing commander there could know what was going on. During that time period it was always rumored by the pilots that that was a cover for in fact the location of one UFO craft."[35]

Note that all he knew was of the rumor; however, it is confirmed by the next testimony, which also confirms some of the comments made by Captain Uhouse.

Mark McCandlish is an accomplished aerospace illustrator and has worked for many of the top aerospace corporations in the United States. A colleague, with whom he studied, has been inside a facility at Norton Air Force Base, where he witnessed alien reproduction vehicles, or ARVs, that were fully operational and hovering. He states that the United States not only has operational antigravity propulsion devices, but we have had them for many, many years, and they have been developed through the study, in part, of extraterrestrial vehicles over the past fifty years.

The colleague, Brad Sorensen, told him of visiting the "Big Hangar" during an air show at Norton Air Force Base on November 12, 1988, and how he had seen flying saucers in this hangar. "There were three flying saucers floating off the floor—no cables suspended from the ceiling holding them up, no landing gear underneath—just floating, hovering above the floor. He said that the smallest was somewhat bell-shaped. They were all identical in shape and proportion, except that there were three different sizes. They had little exhibits with a videotape running, showing the smallest of the three vehicles sitting out in the desert, presumably over a dry lakebed, some place like Area 51. It showed this vehicle making three little quick, hopping motions; then [it] accelerated straight up and out of sight, completely disappearing from view in just a couple of seconds—no sound, no sonic boom—nothing.

"Well, this craft was what they called the Alien Reproduction Vehicle; it was also nicknamed the Flux Liner. This antigravity propulsion system—this flying saucer—was one of three that were in this hangar at Norton Air Force Base. [Its] synthetic vision system [used] the same kind of technology as the gun slaving system they have in the Apache helicopter: if [the pilot] wants to look behind him, he can pick a view in that direction, and the cameras slew in pairs. [The pilot] has a little screen in front of his helmet, and it gives him an alternating view. He [also] has a little set of glasses that he wears—in fact, you can actually buy a 3-D viewing system for your video camera now that does this same thing—so when he looks around, he has a perfect 3-D view of the outside, but no windows. So, why do they have no windows? Well, it's probably because the voltages that we're talking about [being] used in this system were probably something between, say, half a million and a million volts of electricity." Brad Sorensen stated that at the ARV display, "a three star general said that these vehicles were capable of doing light speed or better."[36]

All of these witness testimonies do not prove the existence of a successful U.S. covert program in antigravity technologies. Only the demonstration of such craft coupled with the accompanying government and technical specification documents would "prove" this. However, these testimonies, coupled with information from other substantial sources such as Nick Cook's 2002 book mentioned below, strongly support this contention.

THE HUNT FOR ZERO POINT

The Hunt for Zero Point[37] contains some of the strongest evidence yet for major efforts and successes in the field of antigravity technology. The author, Nick Cook, who for the past fifteen years has been the aviation editor and aerospace consultant for *Jane's Defense Weekly*, spent the last ten years collecting information for the book. This included archival research on Nazi Germany's antigravity technology and interviews with top officials at NASA, the Pentagon, and secret defense installations. He shows that America has cracked the gravity code and classified the information at the highest security levels. Because antigravity and its allied zero point energy technologies potentially offer the world a future of unlimited, nonpolluting energy it has been suppressed because of the "huge economic threat." His findings support those reported by many of the Disclosure Project witnesses cited above.

ANTIGRAVITY TECHNOLOGY
DEMONSTRATIONS

Although Brown reported many of his findings nearly a half-century ago, other experimenters have just recently begun to reproduce his work and report on it in the open literature and on the Web. For example, Davenport published the results of his work in 1995 supporting the findings of Brown,[38] while Bahder and Fazi in 2002 described their assessment of the forces associated with an asymmetric capacitor.[39] Transdimensional Technologies[40] in the United States and JLN Labs[41] in France have posted on the web: diagrams, web videos, and data on their versions of antigravity "Lifters" based on an extension of Brown's work. It is a sad commentary on this whole area of research to see that public science is requiring us to demonstrate principles that were demonstrated nearly fifty years ago.

There have also been a number of other demonstrations of antigravity phenomena by researchers throughout the world. This includes the work of Brazilian physics professor Fran De Aquino and such devices as the Searl electrogravity disc, the Podkletnov gravity shield and Project Greenglow, the Zinsser kineto-baric field propulsion, and the

Woodward field thrust experiments on piezoelectrics. All of these are described in more detail by Greer and Loder.[42]

IMPLICATIONS OF THIS RESEARCH

- Antigravity and zero point energy research and their applications are finally being addressed by some of the open scientific community. This means there will have to be a rewriting of textbooks in this area so our new generation of students can apply this "new knowledge." Its application will lead to major breakthroughs in transportation technologies both Earthside and in outer space. The implications are that we have the potential for human exploration of our solar system and beyond, if we have the will, within our lifetimes. It also means that the majority of twentieth-century space technology will be obsolete and may in fact already be so.

- The zero point or vacuum-state energy source is seen as a totally nonpolluting energy source, which has the potential to replace all the fossil fuels on this planet. It also will provide the energy needed for long-range space flights. This means that fuel cells and solar cells in common use today for space-flight energy applications will only be needed until we transition to these new energy technologies.

- Based on an analysis of trends in antigravity research over the last half-century and the information provided by numerous witnesses, it appears there is both good and bad news. The good news is that it appears that we (at least covert projects) have already developed the theories of antigravity and additionally have developed working spacecraft based on these principles. The bad news is that these technologies have been developed for at least several decades, at the public's expense, and that humankind has been deprived of these technologies, while continuing to waste energy using less efficient and pollution-enhancing technologies.

- Supporting this contention is the following quote from Ben Rich, former head of the Lockheed Skunk Works. Just prior to his death, he stated to a small group after a lecture, "We already have the means to travel among the stars, but these technologies are locked up in black projects and it would take an act of God to ever get them out to benefit humanity." He further went on to say, "Anything you can imagine we already know how to do."[43] These are strong words from a knowledgeable deep insider and words that support what a number of the witnesses stated as well.

- As the reality of this knowledge begins to be understood, there will be an outcry among space scientists not on the inside for release of these technologies to allow all of us to explore space. There will be major changes in the way NASA does its business, though predicting these changes is difficult.

- Not only has space exploration in the public sector suffered, but our planet's environment has suffered as well. As this knowledge begins to sink in, there will be an outcry among all concerned citizens on this planet for release of these technologies to allow all of us to reduce and ultimately eliminate global warming and environmental pollution, which so threaten our way of life. These technologies will not only affect space travel technologies, but will also have a profound effect on transportation and energy production on the Earth's surface.

In conclusion, we might consider the observation made by Halton Arp: "We are certainly not at the end of science. Most probably we are just at the beginning!"[44]

A FREE-ENERGY WORLD

I believe we need a completely new macroeconomic paradigm. My vision is that each home and business would have a new-energy-generation system integral to the site and that the energy would be, indeed, free.

The grid—even the proposed "smart" grid—is anything but smart, but is prone to failure, storms, and outages. We need to move to a paradigm where energy is generated where it is used. Once the device is in place, it would be free.

Impoverished areas and peoples should have such systems heavily subsidized or gifted to them. It is for this reason that we should have a charitable trust to ensure that whatever benefits come from any such new invention, that the proceeds go back to benefit humanity and the needy. Anyone not in agreement with this overarching vision is not allowed to be on our board.

My vision is also to engineer such systems so they last at least one hundred years—eliminating the current paradigm of throwaway devices (lightbulbs, cars, computers, appliances, and the like—all deliberately designed to fail in a short time to churn up more replacement revenue).

Eventually, we will move to applications of the technology so that every manufactured item that uses energy will have a free-energy generator within it—this is completely feasible. At that point, homes and buildings will not even need electrical wiring, saving huge amounts in building costs—and lives—by eliminating electrical fires and electrocutions.

The costs of manufacturing, transportation, agriculture, and other energy-intensive industries will plummet. Eventually we will move to a less-than-twenty-hour workweek with full employment and abundant leisure time to pursue health, recreation, education, and spiritual development and higher states of consciousness.

The current economic macroeconomic system is, in effect, undeclared economic slavery and must be abolished, just as actual slavery was abolished in the 1800s.

That's the view as I see it.

Note that currently all the "new-energy" proposals so much in fashion now are, in fact, a fraud: centralized utilities using multibillion-dollar wind, solar, and similar old technologies—all so it can be *controlled and metered* by the GEs of this world. Absurd.

Of course, decentralizing the macroeconomy and creating abundance for all who dwell on Earth is precisely what the "controllers" wish to avoid: they *like* being masters of the universe. But it is time for change. And it will be the greatest change in known, recorded human history.

THE UNIVERSE
AS ENERGY SOURCE

Energy is a snowflake, a flower, a leaf. It is atoms and molecules, DNA and all the rest. It is the giant redwood and the gentle violet. It is ice-crystals on the window pane, the wonderfully intricate tapestry of tissues and cells in living creatures. It is the cycle of the seasons, the span of a life, the rhythm of the planets. It is the hazy calmness of a summer's day, the power of the hurricane, the crush of a great ocean breaker. It is the gentle flame of a candle, the explosion of a supernova. We cannot escape from energy. . . . It is both the information content and the carrier wave. We are embodied in energy. . . . It is this intrinsic oneness within the many which is the essence of causality.

JOHN DAVIDSON

There is really nothing sacrosanct about the physicist's present interpretation of Nature. We are all free to think things out for ourselves and we can explore our own ideals without being obliged to conform to the pattern already set by others. If we are able to fashion the basic structure of Nature we cannot be timid in the approach we take. . . . We should not invent a pattern of scientific behavior and expect Nature to conform. We should perceive Nature's own pattern.

HAROLD ASPDEN

The implosion motor, however, is centripetally operated. It produces its own driving source through the diamagnetic use of water and air. It does not require any other fuel such as coal, oil, uranium or energy derived from atomic splitting, since it can produce its own energy (atomic power) by biological means in unlimited amounts—almost without cost.

OLOF ALEXANDERSSON

Looking at things from first principles, we have to understand that motion is the essence of manifestation. Everything we perceive is an energy dance. At the physical level, this subatomic dance is spun out on the energy of space or the vacuum state energy field. And the nature of the motion of the spending, whirling energy vortices which we call subatomic particles is of the utmost importance, for it is a pattern which gives rise to all the microscopic forms we perceive with our senses and allied instrumentation.

JOHN DAVIDSON

We have no means of getting hold of the ether mechanically; we cannot grasp it or move it in the ordinary way: we can only get it electrically. We are straining the ether when we charge a body with electricity; it tries to recover, it has the power to recoil.

SIR OLIVER LODGE

Orthodox scientists have never looked for a way to engineer the vacuum, because they have not realized that it is composed of pure massless charge.

THOMAS E. BEARDEN

A prerequisite for solving our self-created environmental planetary mess is a world view that must bear some relationship to the highest spiritual reality. Without this, there is no basis upon which to determine our actions, no intuitive comprehension of natural law, no rationale for unselfish behavior.

JOHN DAVIDSON

10
HARNESSING NATURE'S FREE ENERGY

THE SEARL EFFECT GENERATOR

John R. R. Searl and John A. Thomas Jr.

> *I think it is possible to utilize magnetism as an energy-source. But we science idiots cannot do that; this has to come from the outside.*
>
> WERNER HEISENBERG

> *If the incredibly powerful invisible background is the ultimate origin of spin and therefore of magnetism, then a self-running magnetic motor can't be accused of violating any laws of physics.*
>
> JEANE MANNING AND JOEL GARBON

THE SEARL EFFECT: A DEFINITION

An effect based on magnetic fields that generates a continual motion of magnetic rollers around magnetized rings (also called plates), producing electric energy with no apparent input from outside, and, under certain conditions, creates an antigravity effect that can be used for propulsion.

240

Side effects include negative ionization of the surrounding air and a cooling of temperature around the device when in operation.

JOHN SEARL—AN INTRODUCTION
John A. Thomas

The young John Searl met with a hard life. He was taken from his mother, who was deemed unfit to care for him, and placed in an orphanage. From there he was placed in foster homes, where he was physically beaten on numerous occasions. He had no formal education as a child, but during these years he had a series of recurring dreams. These dreams came twice a year in pairs—first one, then the other—always the same dreams. They recurred from the ages of four and one-half to ten. The dreams were frightening and left a lasting impression on him. He had no idea what they meant, if anything. But they were destined to provide the keys to his life's work.

Searl later discovered a mathematical function called the law of the square (referencing numbers, not the shape). He was immediately drawn to this idea, questioning its significance and how it was used. The more he studied, the more he discovered.

BACKGROUND ON THE
LAW OF THE SQUARES

The history of magic or Latin squares dates back thousands of years to the time of the pyramids of ancient Egypt, and possibly back to the first Chinese dynasties. The Chinese still use magic squares to cut and trim bonsai trees to the correct mathematical ratios of nature. When these laws are understood, one can make a case that the universe exists and acts according to precise mathematical laws.

The law of the squares, or magic squares, is an old technology that experienced a rebirth after Searl discovered the matrix has three-dimensional properties that can be used to model the quantum energy states of mass in space-time.

There are three groups of squares, and there can be no others. Group one squares consist of all odd numbers. Group two is composed of even numbers divisible by four. The remaining even numbers not divisible by four are in group three.

When the correct matrix of random numbers produces the same line value when summed up vertically, horizontally, and diagonally, then you have the law of the squares—a law as valid as the law of conservation in physics, which says energy can be neither created nor destroyed but can be converted from one state to another.

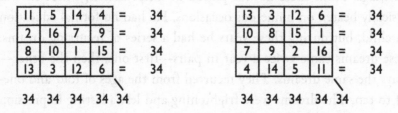

Figure 10.1. An example of the law of the squares.

Searl has written many books on the subject, including *The Law of the Squares,* and has used these mathematical matrixes and their laws to construct his Searl Effect Generator (SEG).

For an SEG to work effectively, the various magnetic elements must be of a specific nature, weight, and dimension, and must be configured in a way that meets the requirements of the law of the squares.

THE LAW OF THE SQUARES
John R. R. Searl

The law of the squares states that all things in the universe have two prime states—opposing states with one having a greater value than the other—and that a division separates these two states. In my books are hundreds of examples, but let us here take the case of *Homo sapiens*— you and me—and see how the law of the squares applies:

1. Nonexperts represent the higher value.
2. Experts represent the lower value.

3. Between these two states there is a thin dividing line of inventors.

4. The nonexperts live in the present, not the yesterdays or the tomorrows.

5. The experts do not live in the present but the past, be it the past of five or a thousand years ago.

6. The inventors live in the tomorrow, for they create the future that is to be. Without inventors there would be no tomorrows, merely a continuation of todays, unchanging except for a change of seasons.

Inventors are accused of living in a world of fantasy, while so-called experts claim to live in the world of reality—ironically, the very world born of the fantasy of inventors who suffered and struggled to create the present reality that the experts now claim as their own.

Before anyone can conceive an idea leading to a new invention, a number of triggers have to be in place:

1. The mind must be free of all conventional brainwashing so the subconscious can communicate freely with the conscious mind. The data stored between these two states can then be activated in response to the correct input or sensory stimulus.

2. The brain must also possess a wealth of information either brought in at birth or acquired over time.

3. Once the correct triggers have been stimulated so the conscious or subconscious mind can start searching and assembling a dream or thought pattern, data make their way into the conscious mind and impress the inventor with the possibilities of some new creation. In the words of Louis Pasteur, "Fortune favors the prepared mind."

4. Wholly original creations are rare because all inventors rely on discoveries and inventions of the past, a past often extending back into the night of time.

5. My work is based on the success of hundreds of inventors before me; my work merely extends their work. Others invented the materials, tools, and equipment that make it possible for me to create.

6. Within nature's laws nothing is impossible except as the mind makes it so. It is the human mind—yours and mine—that has delayed and still delays progress toward a better world than the one we have today.

7. In my case, the subconscious assembled an extremely simple game that strangely contains the mathematics required to achieve the technology that led to the creation of the Searl Effect Generator and the Inverse-Gravitational Vehicle.

8. I was blessed with these discoveries, which contain so much information on science and technology. Like Austrian water pioneer Viktor Schauberger, lack of formal education preserved my native curiosity.

9. The device is basically a linear motor riding on a magnetic bearing; in reality there are three such units in one body.

10. Its characteristics are similar to that of an autotransformer— again, three such devices in one body.

11. The inner unit generates its own output, which feeds into the second unit, which also generates its own much higher output, vastly boosted by the input of the first unit. The second unit's output then feeds into the third unit, which is also generating a much larger output, again greatly increased by the input from the second unit.

12. Every operation of the SEG follows Newton's laws precisely. Experts who claim my work is "impossible" are merely protecting the status quo and their own egos. Ever since Cremonini and Libri refused to look through Galileo's telescope, the experts of each generation have resisted new discoveries. It is the same story today.

13. Not only is there a problem with designing and testing the specialized equipment needed for this research, but opposition also comes from those who function out of ignorance or greed. To seek new solutions by following the same old tried and failed paths is what we call neurotic behavior. But for these frailties of human nature, my technology would have reached the marketplace as far back as 1968 or again in 2003.

14. Today, for the fifth time, efforts are being made to move this technology, but cost is the greatest problem and increases each

year it takes to reach the marketplace. There is also another problem affecting delay, and that is that I have to live and maintain a job in order to survive while covering my developmental costs out of pocket. These problems lead to time delays, but I am absolutely dedicated to the survival and triumph of the SEG, with its enormous benefits for humanity.

15. This world is dying from pollution for which this technology offers a solution no other known technology presently offers. Yet this technology is ignored because it poses a challenge to the immense wealth tied up in existing fossil fuel and related industries. The only thing keeping my SEG out of the marketplace is greed, while humanity suffers and the planet, with its many magnificent species, heads toward extinction.

16. Were a technology such as my SEG, with its capacity to extract energy from the creative vacuum, in the marketplace, it would slowly bring an end to hurricanes such as those experienced along the coastline of the United States, which will also reach Canada and even the United Kingdom if global warming is allowed to increase.

17. You may wonder if the SEG output is direct current or alternating current. It is an alternating current of very high voltage that must be passed through a step-down transformer for the voltage required, like all other power systems available today; current is proportional to the voltage.

18. Rare earth elements play an important part in the success of the SEG's operation. The one that has become the keystone of my work is neodymium (Nd), because everyone wants to follow the same one I originally used in 1946. It is far from the best, but it is the first choice for cost and availability.

19. One major problem is that it loses electrons to the air. As such it will oxidize in air. This can be prevented by the following means:
   a. Keep it in argon while not in use, which is the better choice, or in oil, which is the choice I do not prefer.
   b. Or mix in another rare earth element that does not lose electrons to the air; that will block the Nd from losing electrons.

c. As you move up the element chart there are better options to use. Also it makes the SEG smaller for the same output.

20. Experts have always declared humanity's advances "impossible"— electricity, the automobile, the airplane, the telephone, television, space travel. The great advances of the world are always brought about by men and women of vision who live in a world of fantasy, for they alone are capable of the free thinking by which progress is achieved and humanity's knowledge is advanced.

21. All human achievements originate in the world of fantasy and existed there before making their appearance on Earth, and all inventions of the future will begin there as well. But before any new invention can be brought forth from the world of fantasy into the world of reality, some ingenious person must attune to it. That action requires a set of triggers in a precise order of events. Then and only then will such products become available to mankind.

22. Though the SEG is a marvelous device that can help reduce pollution problems and global warming and give plants and animals a chance to survive, it first requires everyone to care and help clean up this planet that we've made such a mess of due to our lack of knowledge and understanding. It requires that we replant the trees that in our ignorance we have cut down. Otherwise hurricanes and tornados will only increase. It takes about thirty years for a tree to become fully functioning in its duty cycle in reducing global warming. Clearly, we need to begin a massive replacement task to rebalance nature to our advantage.

23. If you truly care about your children's health and welfare, you must act today and help solve these problems. I invite you to assist me in my efforts to get this SEG technology out into the world. This isn't about me. *It's about saving our planet before it's too late!*

THE SEARL EFFECT TECHNOLOGY
John A. Thomas

The Searl Effect Technology promises to be one of mankind's greatest inventions and a door into a new world of boundless possibilities. It

Figure 10.2. Searl technology has all the hallmarks of a world-class solution for today's energy needs. We expect it to be one of the dominant forms of energy production of the future.

could place Searl in the annals of history, but his one desire is to see his technology fully implemented as soon as possible for the benefit of the Earth's people and environment.

The story of his hardships and persecution is a long one. Long after most men would have given in to the gambit of the big business monopolies, the misleading government officials, and the pathological skeptics sponsored to promote disinformation against alternative energy solutions, Searl steadfastly remains with us today, courageously willing to offer his solution to our planet in crisis.

The SEG is a revolutionary device that generates electricity with no apparent input from the outside and is capable of supplying electrical power. It produces economically sustainable clean energy with no reliance on fossil fuels such as oil or natural gas. However, the development and implementation of this revolutionary invention has encountered a great deal of resistance over many years because of the obvious economic repercussions to big oil industries and to centralized energy distribution monopolies that have entrenched themselves deeply within governments and industry.

If the people of the world should choose to support Searl's final mission and adopt his technology, we could eliminate air pollution, solve our looming energy crisis, and support unprecedented economic growth. Anything that can be run electrically can be driven by an SEG with no pollution and no use of fuel as commonly known.

How much longer will we allow the corruption and greed of the privileged few to blind us to the opportunity to heal our Mother Earth and advance us technologically? We hear about massive destruction of tropical rain forests and growing deserts around the world. We also hear about pollution going into our air, rivers, lakes, and seas. Now, the polar ice caps are receding, with a growing ozone hole over them. Also, 2012 proved to be the warmest year on record since global temperature records began being kept. All this is making it increasingly self-evident that the environment is being compromised ominously by human activity. This may be our last chance to make a transition to a better world.

These ecological problems may seem far from our everyday life as most things may still seem normal to us on a day-by-day basis. But we might just wake up one day to the fact that we are too late to repair the damage done to our environment.

THE BASIC STRUCTURE OF A SEARL EFFECT GENERATOR

An SEG consists of a series of three rings and rollers that go around those rings. The first ring contains twelve rollers; this is equivalent to having a twelve-cycle or twelve-phase linear motor. A linear motor will not operate on less than twelve phases. There are many other correlations one can make, but this works along with the laws of Nature. There can be more than twelve rollers, but no less. Each roller consists of eight separate stacked segments held together magnetically.

The SEG is made using Square 4. You must pick a line of Square 4 that you want to use. The numbers of each line add up to the same total. Choose one of these lines to use for building. The numbers represent the weight of each of the elements used. Square 4 can be raised to your desired mass and weight by raising the level of the Square 4. The final number must be that of weight.

The rollers revolve around the plates that form the rings, but they do not touch them.

There is a center element and then three other elements going out from the central core of each roller. The plates will have the same center

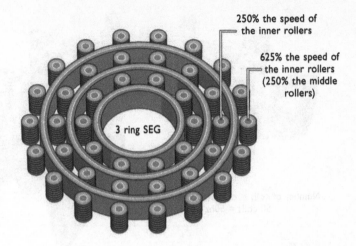

Figure 10.3. In an SEG, the number of rollers on each ring is different—the number increases by ten rollers for each successive ring. All of the construction details are based on the law of the squares.

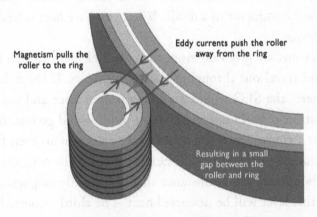

Figure 10.4. Forces acting on the roller

element on the inner side and the lightest element on the outside.

The amount and weight of each element is determined by the level of the Square 4 used. The numbers of the square represent the weight of each element. The largest number is the rare earth element. The second highest number is the magnetic layer. The third highest number is the conductive layer, and the smallest number is nylon 66.

Nylon 66 is the element Searl chose to use because it has a high

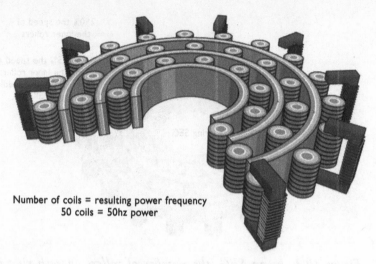

Number of coils = resulting power frequency
50 coils = 50hz power

Figure 10.5. Coils collect energy

negative content and also has a double bond configuration. The nylon is used as a semiconductor in a diode. What you have here is basically a solid state device.

The electrons are given off from the center element (which is neodymium) and travel out through the other elements. If the nylon had not been there, the SEG would act like a pulsed laser and one pulse would go out, stop, build up, and another pulse would go out. But the nylon acts as a control gate; the control gate gives you an even flow of electrons throughout the SEG. The next layer out is the magnetic layer. This layer should contain a substance that will readily magnetize. The specifics of this layer will be discussed next. The third or outer layer is the conductive layer. This is the discharge point of the generator, and the electrons are picked up by C coils around the generator. The output of the generator is high voltage AC, which is then transformed to suit your needs.

THE IMPRESSED MAGNETIC FIELDS

The rollers have a primary north and south pole, as do the plates (large rings) in this case. There's a primary north and south pole on the rollers and a primary north and south pole on the plates. Obviously the north pole of the roller will be attracted to the south pole of the plate.

Ordinarily they would clamp right on and not be able to move, especially being made with neodymium. But there's a secondary field impressed on these magnets using an AC component. By impressing an AC magnetic field on the roller, Searl was able to create the demonstration bar and two rollers. The rollers would rotate completely around the bar, even around a 90-degree angle, because of the way the fields were impressed on the magnets.

One of the marvelous things about the SEG is that none of its parts touch. There is no friction whatsoever. The rollers float on the magnetic field because of the AC component impressed on them and will not fly off because of the DC component impressed on them. They travel around the first ring at about 250 miles per hour. In each successive ring moving outward, the rollers' speed increases by two and one-half times.

ELEMENT SELECTION

The choosing of the elements is of primary importance, obviously. If one looks at the periodic table of elements, one notices that the atomic structure of each element is illustrated. Searl discovered that most of the elements that work best are hexagonal, excluding iron. He uses them in conjunction with the hexagonal configuration in order to pick the correct and most efficient way to develop the desired effect.

The specific powders of elements are carefully weighed. These elements have to be of a specific grain size, atomic weight, and exact dimensions and put together in a way that meets the requirements of the law of squares.

These elements are chosen according to the law of the squares and employed so that the heaviest atomic element is at the center and the atomic weights lessen as the elements progress outward. The center element Searl has used since his first SEG is neodymium, which is element number 60. Modern science has just discovered that neodymium is good for making magnets. Searl has been using it since 1946!

RECOVERING THE DREAM

Various people in history have claimed to have generated what the scientific community calls free energy. And many of them have claimed to have produced antigravity devices and craft. Today, most of them are dead. We're trying to retrace their endeavors to rediscover their technology. Most of these inventors were very secretive to prevent their inventions being stolen or falling into the wrong hands. Their motives are understandable. The result is that their technology and inventions were lost.

These great inventors include Nikola Tesla, John Keely, Viktor Schauberger, Otis Carr, and perhaps others of whom I'm unaware. I come across people almost daily who say they have free-energy devices that work. Searl has kept many aspects of his technology secret for the same reasons Keely, Tesla, and others did. He has been wrongfully imprisoned and had all of his equipment and papers destroyed, almost burying the Searl technology.

But Searl is not to be stopped. He is strong willed and dedicated to using his technology to create a better world for humanity. He has released this work in the form of books titled *The Law of the Squares* so it cannot be destroyed again.

I've asked people, What would you give to be able to talk to Nikola Tesla or John Worrell Keely? I'd give a lot to talk to Tesla, Keely, or many others. What marvelous new inventions might they come up with in light of our modern discoveries? Take Tesla today and show him the transistor; what do you think he'd come up with given his past demonstrations of inventive and visionary powers?

Did Keely really conquer gravity using sound waves through sympathetic vibrations? Did Schauberger really create a gravity field using vortexical action in liquids? Did Carr have special windings in his craft that allowed him to generate voltage needed for antigravity? These questions may never be answered except by diligent research and experimentation, and even then it may take years.

Searl is alive today and wants to give his technology to the world. Will we wait too long to accept it? Will we waste our time scoffing and trying to disprove his technology, as has been done with others in the past? Will we turn a blind eye to his attempt to move us into the future, as we did with Tesla—all so we can hold on to the past?

I say we must not—dare not—turn blind eyes. We're looking for the visionaries of today to move us ahead while there is still time to save the people of this planet. We no longer have the luxury of saying, "Someday someone else will do it." We've severely damaged our atmosphere. We're depleting our natural resources. We're cutting down our rain forests at an alarming rate. We hear about these things and feel we can do nothing, but *real answers do exist,* here and now.

We can do something about these problems if we choose to act before it's too late and the technology is lost once again. When we look at the big picture of the Earth and what we've done to our atmosphere, soil, and water, we realize there is only a minute amount of time left before action comes too late. Some cities are almost unlivable now because of smog. Searl's technology could do something about that. And that's why I'm involved with it.

11

COLD FUSION

THE END TO CONVENTIONAL ENERGY AND THE START OF SOCIAL REORGANIZATION

Edmund Storms, Ph.D.

> *The field of study called cold fusion was born de novo. It did not emerge from a recognized body of continuing scientific research. It was not an extension of ongoing scholarship. No precursors puzzled the world's scientific laboratories. More dramatically, it threatened the canons of nuclear physics. This birth will prove unique in the annals of science.*
>
> CHARLES G. BEAUDETTE

> *It is really quite amazing by what margins competent but conservative scientists and engineers can miss the mark, when they start with the preconceived idea that what they are investigating is impossible. When this happens, the most well-informed men become blinded by their prejudices and are unable to see what lies directly ahead of them.*
>
> ARTHUR C. CLARKE

> *One of the problems the fusioneers have had is that their experiments last far longer than the media's attention span.*
>
> THE ECONOMIST, SEPTEMBER 30, 1989

WHAT MAKES THE SAGA OF COLD FUSION
WORTH YOUR TIME?

How is it possible for one of the most important discoveries of the twentieth century to be systematically rejected by the scientific establishment and some governments for twenty years? This discovery claimed that nuclear reactions could be made to occur in solid materials and these reactions could produce useful heat without harmful radiation. A typical scientist asking this question would answer that the reason is obvious—because the claim is wrong and based on bad science. If bad science were the answer, why would scientists in major laboratories in more than eight countries continue to study the effect? Why would scientific papers continue to be written and published? Why would respected authors write articles and books supporting the claims?

Unfortunately, this type of uninformed rejection is not unique, and many examples have occurred during the past decades and centuries. The present debate about global warming is the most recent example, while the rejection of Galileo's ideas about the sun and planets is a famous example from the past. But we are now living in modern times when new insights are commonplace and when the open-minded ideal of science should be taken for granted—or should it? Perhaps the treatment given to cold fusion shows we were too hasty in thinking science or public awareness had actually changed over the years. Perhaps the average human mind is incapable of objective evaluation of new ideas, even when that person is trained in science. Or has an objective evaluation of Nature failed at all levels, especially in the United States? These questions add to the importance of understanding the history and reality of cold fusion.

For twenty years, a small minority of scientists accepted the possibility, at the risk of their reputations and jobs, that cold fusion was a real phenomenon. What made them different from the skeptics who risked nothing by their rejection? To make matters worse, in many cases simple rejection was not enough. The skeptics also made sure the claim became an allegory for bad science. As a result, if a lazy writer wanted an example of pathological science, he had only to use the claim for cold fusion as an example of how science can go astray. Use of this approach might be forgiven as normal incompetence, except for one important fact.

Cold fusion has the potential to completely reorganize society—an importance that does not tolerate ignorance or incompetence. Therefore, understanding the consequences of this energy source is essential before it gains widespread use. Equally important is understanding why such universal rejection was so widely accepted. Hopefully, similar mistakes might be prevented in the future.

Cold fusion has important benefits as well as some handicaps. Let's start with the advantages. The resulting energy involves no gas emission or radioactive by-products. Once cold fusion is adopted, global warming resulting from atmospheric CO_2 (carbon dioxide) and CH_4 (methane) can be a concern of the past. Even if a reduction in atmospheric CO_2 does not stop global warming, a great deal of energy will be required to build protection for countries that will be exposed to the rising oceans and that will need to rebuild after the damage produced by severe weather we can expect in the future. This energy source will be especially important as conventional sources become increasingly expensive due to growing demand and dwindling supplies.

These are the major advantages, but several problems also exist. The process is not easy to initiate on purpose. Skeptics use this lack of reproducibility to reject the whole idea. This difficulty occurs because an unknown process starts the nuclear reactions in an unknown special material that is created mostly by chance. The challenge now is to identify how this special material is formed and deliberately make large quantities.

Although temperatures as high as 450°C using fused-salt electrolysis[1] have produced excess heat and helium, this success has been difficult to replicate and the lifetime of the process might be short. Temperatures at the boiling point of water[2,3,4] have also produced successful results, but this temperature is not ideal although it can be useful. Until the special material required to initiate the nuclear effect can be made in large amounts with reliability, the engineering problems will remain unsolved. Nevertheless, the advantages and need are so great that this phenomenon is worth exploring in spite of a few potential problems. A rational person has to ask, "With the worldwide interest in finding energy sources that can stop global warming so intense, why is this obvious potential energy source ignored?"

A BRIEF HISTORY OF COLD FUSION

I learned of this discovery while working at the Los Alamos National Laboratory (LANL) in 1989, and I continued to investigate the claim in my own laboratory after retiring from LANL. In addition, I published dozens of papers and reviews, including a book describing all the reported observations up to 2007.[5] What I'm about to describe is not based on incomplete knowledge or popular opinion.

Anyone reading newspapers twenty years ago anywhere in the world will remember Professor Martin Fleischmann (then head of the Department of Electrochemistry at the University of Southampton) and Professor Stanley Pons (then chairman of the Chemistry Department at the University of Utah).[6] In March of 1989, these two chemists announced an incredible discovery. They claimed to cause a heat-producing nuclear reaction using conditions considered to be impossible. Nevertheless, initially the media promoted the idea of cheap energy made by a process so simple any high-school graduate could make it work. The energy needs of the world would be solved by a process called cold fusion, so the story went.

This energy was claimed to result from two deuterons, an isotope of hydrogen present in all water, fusing to produce a huge amount of energy and several other elements, all in a simple electrolytic cell. The expensive, long-promised, and largely unsuccessful effort using high-temperature plasma, called hot fusion, would no longer be necessary. The use of oil and coal, sources of pollution and environmental problems, would be a concern for history to evaluate. Initially, these claims were exciting, but they quickly became too hard for many people to believe when replication turned out not to be as easy as promised and the consequences of the claims sank in. The threat to the hot-fusion program and the oil industry became all too obvious. Besides, no one could explain how such a process could actually work. These two issues combined to produce a perfect storm of rejection.

Very soon after the announcement, several books were written describing the whole idea as pathological science. These books laid the groundwork for rejection, along with help from several respected scientific publications, as the following examples show. Gary Taubes[7] is a science writer who was intent on promoting his hopefully popular book called *Bad Science: The Short Life and Weird Times of Cold Fusion.* To help the process, he created a totally unsupported scandal about data being falsified at Texas A&M University.[8] The publication *Science* supported this spurious accusation without apologies or explanation, even after the claim was shown to be false.[9] As a result, reputations were damaged while misinformation spread to a growing number of scientists. John Huizenga, a respected scientist chosen by the U.S. government to evaluate the subject, showed his skepticism in his book, *Cold Fusion: The Scientific Fiasco of the Century,* published in 1992.[10] The well-known scientific journal *Nature* helped promote this attitude, thanks to its editor at the time, John Maddox.[11] Frank Close, another respected scientist, added to the climate of rejection with the book *Too Hot to Handle.*[12] In contrast, a balanced and honest book by Eugene Mallove titled *Fire from Ice: Searching for the Truth behind the Cold Fusion Furor* was largely ignored.[13] Many more books[14-24] have been written over the years, many supporting the original claims for unusual nuclear reactions while taking different points of view. The Discovery Channel, the History Channel, and *60 Minutes* produced accurate and supportive program segments. Gradually, popular opinion is moving in favor of cold fusion, but not before twenty years was wasted in active rejection.

Influential individuals also played important roles. Douglas Morrison[25] added to the general rejection by writing a regular newsletter from his post at CERN (European Organization for Nuclear Research). He maintained his negative attitude until his death, even though he attended conferences about the subject and talked to many researchers. To his credit, he gave an informed evaluation, but one based on a deeply held certainty that a claim for such a nuclear reaction was completely wrong. In contrast, Robert Park apparently made no effort to understand the subject and occasionally rejects the idea by ridicule in his popular weekly newsletter.[26,27] Nathan Lewis, a professor at the California Institute of Technology at the time, used a flawed and failed attempt at replication to fan the growing antagonism among physicists at an early American Physics Society meeting in Baltimore.[28]

This set the stage for rejection by the general physics community. Other less visible individuals took turns rejecting the idea, a process that continues to this day, but with less frequency. I mention these few examples to show how hubris and the arrogant certainty of certain individuals can influence the view of reality accepted by a large number of people—even scientists. This is not to say that skepticism is unimportant. Some claims and ideas are worthy of rejection. It's the intensity of the rejection that causes trouble, especially when such enthusiasm is based largely on ignorance. Such ignorance concerning cold fusion is no longer excusable thanks to accurate information now available on the Internet. (Please see this chapter's listings in the Additional Resources section for more information.)

To my knowledge, none of the people who were publicly outspoken in their rejection of the claims have changed their minds based on the new and more complete information. Skepticism seems to be a one-way process that never admits of being wrong.

Much skepticism was fueled by a lack of theory to explain the observations and an apparent conflict with accepted theories. Apparently, professional scientists need to be reminded that theory should not be substituted for reality, only used as a temporary and imperfect map. A hundred years hence, theories we presently take for granted will look simpleminded—just as theories accepted a hundred years ago now look to us. All new ideas may seem to be improbable when based on currently accepted models, but such is the nature of new ideas. People forget that our presently accepted ideas were once new and initially rejected. The solution is patience rather than dogmatic rejection when new ideas are proposed.

The debate ended as far as the U.S. government was concerned, in 1990, with a widely circulated report from the Energy Research Advisory Panel.[29] Its rush to judgment was made before the claims could be properly explored. Even worse, the group was cochaired by

John Huizenga, whose bias was later revealed in his book. The other cochair, John Ramsey, forced adoption of a more objective conclusion only by threatening to resign. Nevertheless, cold fusion was officially accepted as nonsense. From that time, no government or corporate support for research would be forthcoming in the United States. All patents that went through the nuclear division of the U.S. patent office were routinely rejected regardless of objective evidence.

Nevertheless, patents were obtainable in other countries. Sorting out the many applications will be a major headache once our government accepts the phenomenon. Eneco, a company created to advance the field, spent over 1 million dollars in a futile effort to have the original Fleischmann-Pons patent and other patents granted in the United States. Partly as a result, the company recently went bankrupt, with a total loss to its courageous investors. This rejection by the U.S. government continued until recent times, with a few exceptions. The Department of Energy even made a second review in 2004 and came to the same but less extreme position.[30,31] In contrast, the U.S. Navy and the Defense Advanced Research Projects Agency (DARPA) have both shown independent judgment. Only recently has the U.S. government shown modest interest even as work in other countries has accelerated. Once again, we need to ask why the U.S. government would reject a potentially ideal energy source whose benefits to its citizens could be huge. Perhaps a lack of benefit to the oil and coal industries suggests an answer.

Scientific journals are essential to science because they allow a free flow of information, the lifeblood of advances in knowledge. Most scientists will not accept information that has not been reviewed by their peers and made available in one of these sources. Presumably, they don't view themselves as capable of determining whether information is believable without such certification. Most conventional journals would not accept a paper supporting cold fusion because the editors and peer reviewers thought "the evidence is not compelling," as physicists like to say. A few reviewers were less polite. Of course, papers showing why the effect could not possibly work or that showed a failure to replicate results were published. One important exception was *Fusion Technology* under the editorship of George Miley, of the University of Illinois, who persisted despite many efforts to have him fired because of his open-minded approach to the subject.

This lack of access to most scientific journals stopped new information from changing the initial negative impression about cold fusion. As a result, popular opinion split into two camps—those who accepted the initial negative response and never looked back and those who continued to do research and share information. If a person was identified as a believer, no evidence from that person would be accepted by a skeptic. Apparently, believers were considered too irrational and blinded by their beliefs to be taken seriously by a normal person. Consequently, the "believers" used the Internet to communicate and held regular conferences in various countries. These conferences began in Salt Lake City in 1990 with the First International Conference on Cold Fusion (ICCF-1). Each subsequent conference was held in a different city and often in a different country.

The Fifteenth International Conference on Condensed Matter Nuclear Science (ICCF-15) was held in Rome, Italy, in 2009, under the sponsorship of the Italian scientific societies and the Italian National Agency for New Technologies (ENEA), a government run laboratory. Most of the published proceedings are available and contain a treasure trove of information. A few other conferences on the subject are regularly held in Italy, Russia, and Japan, independent of the ICCF series. These meetings allow exchanges of information and a way for papers to be made available in published proceedings without interference from the general negative attitude. Recently, papers have been presented at regular American Physical Society and American Chemical Society conferences and have been published in the American Chemical Society and American Institute of Physics symposium series. An online peer-reviewed journal called *Journal of Condensed Matter Nuclear Science,* with Jean-Paul Biberian of the Université de la Méditerranée in France as editor, is currently accepting papers on the subject. As a result, facts are gradually seeping into the consciousness of mainstream science.

A few countries did not stop exploring cold fusion after the general rejection, and some individuals in the United States continued research in private or university laboratories using their own money or support from a few wealthy friends. The Japanese government supported a large program, New Hydrogen Energy, designed to determine whether the effect had commercial application, and Technova in Japan supported ongoing work by Fleischmann and Pons in France. Later, Japanese companies supported important work by Michael McKubre at SRI International in the United States when other resources failed. Although these efforts eventually terminated, the information and experience gained fertilized ongoing work at several universities and companies in Japan. Work in Italy also gradually grew into a large program supported by the major scientific societies. Studies are also underway and growing in China and Russia. Even the U.S. government is showing a long-delayed interest. Unfortunately, this increased awareness is coming at a time when financial support for new ideas, however important, is harder to find. For example, although a company in Israel has reported success in making energy with increased reproducibility, it has recently lost its funding.[32]

HOW DOES FUSION WORK?

What is this strange effect that has caused so much controversy? An answer requires some background and technical information I will try to provide. Nuclear interaction between nuclei is prevented because all nuclei repel each other, thanks to each having a positive charge. This creates what is called the Coulomb barrier to direct nuclei interaction. The larger the atomic number, the greater is the charge and the barrier. The three isotopes of hydrogen (protium, deuterium, and tritium) have the smallest barrier, each with a charge of 1. This relatively small barrier gives hydrogen nuclei an advantage in entering into nuclear reactions with each other or with more highly charged nuclei. Reactions can be made to occur between nuclei if they are caused to smack into each other with enough force. This force can be applied either by heating them to a very high temperature in plasma (over 1 million degrees) or accelerating them in various ways. This method is called hot fusion. The resulting reaction generates tritium and neutrons, both of which are dangerous and make the apparatus radioactive, hence expensive to repair.

In contrast, cold fusion uses solid materials containing deuterium as the environment in which the nuclear reactions can occur. This allows fusion and other nuclear reactions to be initiated at room temperature without extra energy being necessary. However, special solid materials are needed that, unfortunately, are created at the present time by lucky chance. Work is underway to identify these materials and purposefully make them in large amounts. This method generates mostly helium, identical to the gas used in balloons, without significant radioactivity. Furthermore, the process is easy and safe to study, an advantage I personally appreciate.

WHAT IS THE STATUS OF HOT FUSION AND OTHER ENERGY SOURCES?

The conventional method, called hot fusion, has been explored for the last sixty years and has consumed at least 25 billion dollars in 2009 dollars, without producing any useful energy. Lots of energy is generated, but not enough to run the huge machine and have any left over. In fact, even if current problems were solved, the machine probably would be too complex and expensive to produce practical power. A picture of the last version, scheduled for construction in France, is shown in figure 11.1. Even this research tool is proving to be too expensive, leaving further construction in doubt.[33,34] For comparison, a photo of Fleischmann and Pons holding one of their cold fusion cells is shown in figure 11.2. Both of these devices are used only for research, requiring much larger versions, in both cases, to generate commercial power. Nevertheless, at this stage, the study of cold fusion is obviously much less expensive than further exploration of hot fusion.

Unfortunately, without this or a similar substitute for conventional but limited energy sources, future civilization is doomed to a much lower standard of living as these sources run out. This realization encouraged the study of hot fusion over the years despite repeated failure to live up to its promise. Time has now run out, and

Figure 11.1. One version of the International Thermonuclear Experimental Reactor (ITER) hot fusion energy generator. Note the size of a person (circled). Additional large equipment required to run the machine is not shown.

Figure 11.2. Stanley Pons (left) and Martin Fleischmann (right), with their cold fusion cells.

we can no longer wait for hot fusion to succeed. It is now time for Plan B.

For many people Plan B is large-scale development of solar and wind power. These sources are quite practical in certain locations, but the power density is low, requiring the use of large tracts of land, sometimes well away from where the power is needed. In addition, they are not constant sources, frequently requiring backup systems to be constructed at additional expense. A complex and sometimes fragile power distribution system is also required, called the grid. A more concentrated and reliable source of power is needed. Conventional fission reactors are such a source, but these are based on splitting uranium into two smaller parts. Uranium has several disadvantages: the supply is limited, mining produces local environmental deterioration, the nuclear reactors produce dangerous radioactive products requiring long-term storage, and the technology has a small but important risk of failure. Furthermore, such reactors can make plutonium for use in nuclear weapons. Clearly, a better Plan B is needed.

WHAT ARE THE BENEFITS OF COLD FUSION?

Cold fusion has the potential to provide energy for many centuries without taking over the landscape, as is required for solar or wind power, and without the environmental damage caused by uranium- or carbon-based power. It is safe and cannot be used directly as a weapon. In fact, it is so safe an energy generator could be placed in each home, eliminating the need for our present complex electrical grid system with its susceptibility to storm-related and other power failures. No more blackouts! The active component, which is deuterium, can be extracted from any water source very simply and without harmful consequences. Heavy water (D_2O), which is the form of water containing deuterium, is presently extracted from ordinary water in large amounts and used as a cooling fluid in certain fission reactors. This liquid can be purchased from any chemical supply company for about fifty cents per gram in small amounts and for much less in large amounts. It is not dangerous, radioactive, or controlled by the government, as is uranium. In short, the source of this energy is as benign as a cup of tea.

HOW IS COLD FUSION PRODUCED?

Four methods have been found to initiate the fusion reaction using very simple apparatus. The original method explored by Fleischmann and Pons used electrolysis to implant deuterium into a palladium cathode, with many variations of this method being explored by other people. In this case, electrolysis involves passing a current through heavy water between two electrodes, one of which is made of palladium. Deuterium is released at the palladium electrode, where it reacts with the metal to form a palladium-deuterium compound. When the quantity of deuterium becomes sufficiently large, the nuclear reaction starts. Achieving the very high deuterium concentration is the main problem in getting the process to work. A typical cell used to measure power production is shown in figure 11.3.

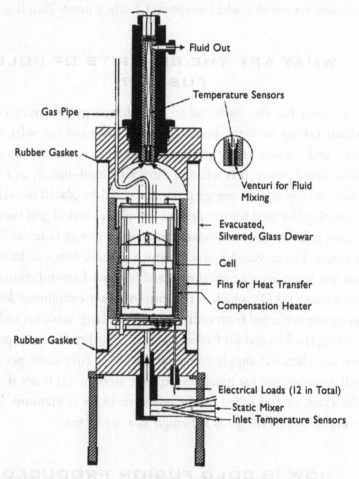

Figure 11.3. Drawing of a calorimeter cell used by Michael McKubre and colleagues at SRI International. The size is similar to the device held by Stanley Pons in figure 11.1.

Most of the time nothing special happens, but occasionally significant extra heat is produced when special palladium is used. This power frequently lasts long enough to accumulate more energy than could result from any known chemical reaction. Hundreds of examples have now been published, some giving power at over one hundred watts from a sample of palladium no larger than a dime in area and total weight. Eventually, this heat was found to relate almost exactly to the amount of helium expected to be produced by the d + d = He fusion reaction. In addition, radiation of various kinds is detected that can only result from nuclear reactions and cannot be mistaken as error. Fortunately, this radiation cannot escape from the device. Work is presently underway to fully characterize this radiation in order to identify the nuclear process.

Work in Russia by Alexander Karabut and coworkers focused on creating a discharge in deuterium gas using no more than a few thousand volts, much too low to produce significant hot fusion.[35] A nuclear reaction is created on the cathode, resulting in extra heat, helium, and modest radiation. Tritium, a radioactive isotope of hydrogen, can also be produced this way, as discovered by Thomas Claytor and coworkers at LANL, when special cathode material is used.[36] This method is more efficient than hot fusion because it uses much less energy to initiate the fusion process and it results in a benign collection of nuclear products. An important advantage is the absence of neutrons, which are dangerous and inconvenient. In other words, the brute force method of using very high temperatures in plasma is not required when the right conditions are created using a special metal after it is reacted with deuterium.

We have an interesting situation. Great effort is made during hot fusion to isolate the plasma from the metal walls using complex and very energy-intensive magnetic fields. Otherwise the walls would be destroyed and the plasma would be degraded. Both the plasma and the magnetic fields use massive amounts of energy supplied by a complex machine. In contrast, after implanting deuterium in a special solid at room temperature, cold fusion achieves the same effect more efficiently using a simpler machine and produces no radioactive by-product. In other words, hot fusion avoids the involvement of a solid while cold fusion welcomes its use.

Other methods have greater potential for producing commercial power. Deuterium can also be implanted in a solid using sound. When intense sound waves pass through a liquid, they generate bubbles that repeatedly collapse and reform. When a bubble collapses on a metal surface, a plasma of ions is injected into the metal at each collapse cycle. This injection of ions produces effects similar to those produced when ions are injected by electrolysis or gas discharge. Rodger W. Stringham has explored this method and found evidence for extra heat and helium production when the liquid is heavy water.[37] This method is in sharp contrast to the difficult and seldom successful sonofusion or sonoluminescence method. In that method, energy is concentrated within a bubble as it collapses within a liquid, resulting in a burst of light just before it disappears. Various people tried to produce hot fusion this way and succeeded mainly in creating a controversy.[38-40] Once again, the attempt made to use high energy—brute force—in plasma produces nothing of value. Once again, injecting deuterium into a solid worked much better.

Finally, a method discovered in Japan by Professor Yoshiaki Arata has potential to produce commercial power with the greatest ease.[41,42] He found that nano-sized palladium powder or certain alloys of palladium can initiate the fusion reaction if the powder is simply exposed to pressurized deuterium gas. Unfortunately, only a fraction of the material has this ability. The challenge now is to find out where in the sample this special material is located and determine its characteristics. In fact, such finely divided material can be identified as being present during all successful occasions when each of the methods is used. This realization provides one more insight into what makes such reactions possible.

In addition to fusion, other nuclear reactions, called transmutation, are found to happen at the same time as fusion. This discovery opens a whole new phenomenon of which hundreds of examples have been published. For example, Yasuhiro Iwamura and his coworkers at Mitsubishi Heavy Industries in Japan were able to convert barium into samarium by adding a cluster of six deuterons to the barium nucleus.[43,44] This process took place in alternating layers of palladium and calcium oxide while deuterium was diffusing through the sandwich. The barium had been deposited on the surface, where the transmutation reaction was observed using X-ray analysis and changes in isotopic composition. The method and evidence are too complex to discuss here, but you can be sure this amazing result has been explored very carefully using very modern methods by several other laboratories.

This is not the only example of such reactions. Workers in many laboratories have reported similar transmutation reactions many times using various target nuclei to produce a variety of other elements. An example is shown in figure 11.4. The platinum (Pt) noted on the figure is not a transmutation product because it readily transfers from the platinum anode. In contrast, the other elements are not expected to be present in the cell. These reactions typically create elements heavier than the palladium target (Pd), but on occasion fragments of these elements are also detected, especially iron (Fe) and titanium (Ti), as seen on the figure.

Errors produced by contamination, that is elements that were previously present, and faulty analysis are real but have been carefully evaluated and reduced to a minimum. For hydrogen to enter the nucleus of such highly charged nuclei means an unknown process must hide the

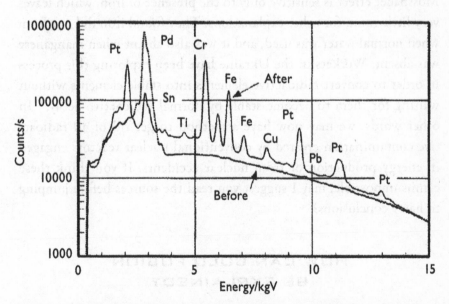

Figure 11.4. Energy-dispersive X-ray spectroscopy (EDX) analysis of a palladium rod before and after electrolysis and production of extra energy

charge on the hydrogen nuclei. Understanding how this can happen is the essence of the problem cold fusion faces on its path to acceptance and application.

For readers predisposed to open-mindedness, you might be interested to know nuclear reactions have also been found to occur in biological life-forms, as first documented by C. Louis Kervran over thirty years ago. His claim was universally rejected, and only now has evidence been published to show he was correct.[45,46] Recently, H. Komaki in Japan studied a variety of single-celled organisms grown in cultures that were missing certain elements needed for life.[47] The living cells were able to form elements they needed to grow using other elements that were present in the culture. These new elements were formed by means of the same type of transmutation reaction found to occur in cold fusion studies. Vladimir Vysotskii and coworkers, in a series of papers[48] and in a book,[49] described how they were able to convert manganese (Mn^{55}) into iron (Fe^{57}) using bacteria grown in heavy water (D_2O) containing manganese sulphate ($MnSO_4$). They used the Mossbauer effect to follow the gradual growth of iron. The Mossbauer effect is sensitive only to the presence of iron, which leaves very little room for other explanations. They found iron did not form when normal water was used, and it was also absent when manganese was absent. Workers in the Ukraine have been exploring this process in order to convert radioactive elements into stable elements without waiting for them to become stable by normal radioactive decay.[50] In other words, we may now have a method to get rid of all radioactive contamination created by conventional nuclear reactors engaged in energy production or from nuclear accidents. If you think these claims impossible, may I suggest you read the sources before jumping to hasty conclusions?

HOW CAN COLD FUSION BE EXPLAINED?

People like explanations, and scientists are especially interested in knowing how something works before they will fully accept its reality.

A wide assortment of explanations for cold fusion have been proposed, far too many to discuss here. Different branches of science take different approaches when evaluating observations. For example, physicists tend to apply mathematical models to ideal materials. When explaining cold fusion, they use the rules and mathematics of quantum mechanics to focus energy on the nuclear process or to modify the relationship between nuclei within a perfect arrangement of atoms. Unfortunately for this approach, the atoms are never arranged perfectly and frequently the environment is a complex mixture of different elements. A few explanations even propose neutrons are involved because this particle can enter a nucleus without experiencing a barrier, thanks to the absence of a nuclear charge. Most explanations are useless because they ignore many observed behaviors and requirements other fields of science impose.

I would like to summarize my approach. Everyone agrees energy is required to overcome the Coulomb barrier. This energy must come from the surrounding environment. This environment consists of electrons and nuclei arranged in a very definite but perhaps imperfect structure. The energy in this environment is being used by the atoms to create this environment and its associated properties, and is not available to do anything else. Any change in this energy would change the atomic arrangement and its properties.

In other words, Nature arranges the atoms in materials with the greatest economy possible, leaving no energy left over. As Henri Louis Le Chatelier observed, Nature tries to undo any change we attempt to make and always uses the least possible energy. Therefore, energy cannot be spontaneously concentrated except in small amounts caused by random variations within the structure. Even if a novel process were to concentrate energy, it would cause chemical and physical changes in the structure before it could reach levels required to initiate a nuclear reaction. Therefore, an energy-concentrating process, or one that rearranges electrons, cannot work no matter how quantum mechanics is applied. This conclusion has to be an essential part of any explanation for cold fusion.

The process must start with the release of energy. This release must involve a process that could occur but is blocked by an energy barrier. Many such processes are present in ordinary materials. For example, many materials will react with oxygen and burn, a process that is prevented by an energy barrier. In the case of cold fusion, the process must be sensitive to a rare condition in a solid and produce a reaction product that is able to initiate a nuclear reaction.

We now know that this process must produce a cluster of deuterons. What process can be proposed to have such a requirement? At this point the explanation requires the existence of a structure not known to science, but hinted at by the properties of the Rydberg molecule[51-53] and the proposed hydrino.[54] Another way of saying this is to state that a two-step process is required. The first step is a chemical one and produces a cluster of deuterons in which the nuclear charge is hidden. This structure is able to cause a nuclear reaction as the second step. Both processes release energy. Over four hundred years ago, alchemists called this unusual nuclear-initiating material the philosopher's stone. Looks like history may once again be repeating itself.

Considerable imagination is being applied to explain these unexpected nuclear reactions. In the process, a new window into understanding nuclear interaction is being once again opened. Perhaps efforts to understand cold fusion will create a whole new field of science. This field is now called condensed matter nuclear science, but the name will no doubt change once again as new insights need to be described more exactly.

WHAT ARE THE DISADVANTAGES OF COLD FUSION?

Why did cold fusion cause such a negative reaction? The rational reason is that this energy source has the potential to be a very disruptive technology. It puts at risk all other sources of energy because of its low cost, simple design, and lack of harmful by-products. Imagine having an energy generator in every home able to provide all the energy needed by the household, including energy for the plug-in auto. The grid would be obsolete, along with the large number of coal-burning power plants. If the transition were not done properly, a lot of people would lose their jobs and even more people would go broke.

Government regulators in the United States and in some other countries would regulate a slow introduction, but what about third-world countries? Many poor countries could quickly have abundant and cheap power available, allowing them to manufacture products at a much lower price. Society would change rapidly, some aspects for the better, while some groups would lose in the process. To make the situation worse, people will apply this new energy source to war. Battlefield lasers would become practical more quickly, with deadly consequences to any army that does not have the technology. Consequently, society had better understand this energy source and apply it with caution because no amount of skepticism or government ignorance will stop the eventual application and the easily foreseen consequences.

CONCLUSIONS

To summarize, an entirely new phenomenon relating a chemical environment to a nuclear reaction has been discovered and rejected prematurely based on self-interest and hubris in some countries and not in others. This process has the potential to become an ideal energy source for centuries. However, the twenty-year delay in exploring the phenomenon leaves many important questions unanswered. Nevertheless, enough is now known to prove the reality of the effect but not enough to know how to make power at commercial levels with reliability. The phenomenon has opened new fields of science and technology with unimaginable consequences.

The issue at this time is not whether cold fusion is real or not—its reality has been established beyond doubt. The problem is one of being able to apply the right tools to discover the nature of the material that initiates the nuclear process. These tools are available, but they are too expensive to be used at the present time without more financial support. Very little progress can be expected until the necessary resources are made available. The progress to civilization that new forms of energy offer outweighs the short-term problems of freeing our planet from the destructive effects of current fuel technologies, and they deserve the funding necessary for their development.

GLOSSARY

Alpha radiation: Emission of a helium nucleus.

Anode: Positive electrode during electrolysis.

Calorimeter: A device for measuring heat energy.

Cathode: Negative electrode during electrolysis.

CERN: The European Organization for Nuclear Research. The acronym comes from the original name in French, which was Conseil Européen pour la Recherche Nucléaire. The group's website is: http://public.web.cern.ch/public.

DARPA: Defense Advanced Research Projects Agency. Agency of the U.S. Department of Defense and responsible for the development of advanced technologies for military use. The agency's website is at: www.darpa.mil.

Deuterium: An isotope of hydrogen containing one proton and one neutron.

EDX: *Energy-dispersive X*-ray spectroscopy. A method of analysis during which a material is bombarded by energetic electrons and the resulting X-ray energy is used to identify the elements present in the surface region.

Electrolysis: A process by which a current is passed through a conductive liquid, during which the liquid is decomposed into its components. In the case of cold fusion, the liquid is heavy water, frequently containing LiOD (deuterated lithium hydroxide), and the heavy water is decomposed into deuterium at the cathode (negative electrode) and oxygen at the anode (positive electrode).

Fission: The process whereby a nucleus splits into several parts.

Fusion: The process whereby two nuclei combine to form new elements. This term is usually only applied to two hydrogen nuclei combining together.

Gamma radiation: Wave radiation emitted from a nucleus as it changes energy.

Hydrogen: Hydrogen is one of the seven diatomic elements, meaning that it is never found alone in nature and always bonds with another hydrogen atom to form H2.

Isotope: The nucleus of an element contains a mixture of protons and neutrons. The number of protons gives the atomic number and determines the kind of element. The number of neutrons plus the number of protons determines the atomic weight. Elements having different numbers of neutrons are called isotopes. Some isotopes are stable and some are radioactive.

ITER: International Thermonuclear Experimental Reactor. International effort to develop hot fusion as a commercial power source. The group's website is: www.iter.org/default.aspx.

Neutron: A neutral radioactive particle consisting of an electron and a proton.

Palladium: An element that reacts well with deuterium or hydrogen to form a compound having the formula PdDx, where x has values between about 0.7 and 1.0.

Plasma: A collection of ions in the form of a gas. The plasma usually has a high energy, denoted as temperature or measured as electron-volts.

Quantum mechanics: A mathematical approach that describes how electrons and nuclei behave as waves. Various accepted mathematical functions are applied to predict the behavior of wave functions based on several different assumptions, including that the structure follows several rules based on quantized energy states.

Tritium: An isotope of hydrogen containing one proton and two neutrons. It is radioactive, emits an electron (beta) for making helium-3, and has a 12.3-year half-life.

X-radiation: Wave radiation emitted by changes in electron energy.

12

ZERO POINT ENERGY
CAN POWER THE
FUTURE

Thomas Valone, Ph.D., P.E.

NOTE: This essay is an excerpt from the book *Zero Point Energy: The Fuel of the Future* by Thomas Valone, Ph.D., P.E.

INTRODUCTION

Up until now the use of zero point energy (ZPE)* for electricity generation was mere fantasy and science fiction. No one thought it possible for ZPE to offer a source of unlimited energy for homes, cars, and space travel. Some "experts" still say zero point energy can do nothing useful. The experimental evidence, however, now incontrovertibly shows that ZPE is present in measurable quantities in coils and semiconductors. Furthermore, ZPE can be turned on and off and produce an attractive or repulsive force in one direction, besides moving electrons to create electricity flow in a circuit and rectified in semiconductor diodes. All

---

*I will use the terms *zero point energy* or *quantum vacuum energy*, while other scientists, including Michio Kaku on the Science Channel, use more obscure terms like *negative energy* or *Van der Waals forces* to mean the same thing.

of these discoveries have been made and published only in the past few years, which indicates that the future energy trend is toward ZPE power.

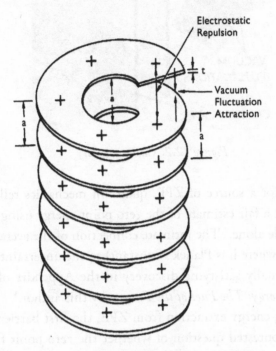

Figure 12.1. Spiral design for a vacuum fluctuation battery

Zero point energy is the sea of energy that pervades all of space and every atom, often called the physical vacuum. Furthermore, it is estimated to exceed nuclear energy densities. But what is it really? It is the nonthermal, "kinetic energy retained by the molecules of a substance at a temperature of absolute zero."[1] Still, most people maintain their doubt as to whether it can be found to be useful for human energy needs. This includes the U.S. Department of Energy "Ask a Scientist" official website. A fifteen-year-old submitted a question, "Would Zero-Point Energy be a better source of power than Antimatter?" The answers include phrases like "a lousy source of energy," "no idea how to do this," "I can't imagine how it could be practical," and "no way to extract the zero point energy from a molecule in order to use it."[2]

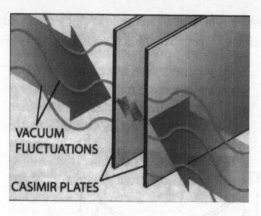

Figure 12.2. Casimir cavity

Looking for a source of ZPE, quantum mechanics tells us, "It is possible to get a fair estimate of the zero point energy using the uncertainty principle alone." The intimate connection of the average value of ZPE (= ½ hf, where h is Planck's constant) to the uncertainty principle is an intellectually satisfying discovery in the Appendix of the book *Zero Point Energy: The Fuel of the Future,* by this author.[3]

Regarding energy extraction from ZPE, the first barrier to resolve is the hotly contested question of whether the zero point field (ZPF) is conservative. (Unlike politics, conservative fields in physics are those that conserve energy, thus obeying the first law of thermodynamics.) If ZPE is not conservative, then we can extract "an infinite amount of energy" from the vacuum, according to the late Dr. Robert Forward. However, if it is proven that the ZPF is a conservative field, we can still extract energy from it. It is just that we would have to put energy in and store it somehow to get it out again, like Forward's invention (figure 12.1). "It does reflect work done by the ZPF on matter."[4]

The evidence so far is in favor of a nonconservative ZPF, with arguments still raging on both sides. In addition, the capability of ZPF storage and retrieval has convincingly been presented by Forward with his ZPE corkscrew charge device he called a vacuum fluctuation battery.[5] An article titled "Energy Unlimited" was published a few years ago when Professor Jordan Maclay from the University of Illinois received a NASA grant to try to extract ZPE from elongated, oscillating, tiny metal boxes.[6] A physics journal article points out, "Vacuum fluctuations remain a matter of debate, mainly because their energy is infinite. More strikingly, their energy per unit volume is infinite."[7] Interestingly, this fact has created intellectual difficulties among physicists that have only been artificially eliminated: "Problems with the infinite energy of vacuum fluctuations has led to the view that vacuum energy may be forced to vanish by definition . . . [from] the need to regularize the infinite energy-momentum tensor." However, this quick fix causes even more complications. "This procedure gives rise to ambiguities and anomalies, which lead to a breakdown at the quantum level of usual symmetry properties of the energy-momentum tensor."[8]

WHY ATOMS DON'T COLLAPSE

To understand how fundamentally real ZPE is for sustaining our reality, it is worthwhile noting that ZPE is fundamentally responsible for atomic stability, much like an "atomic Casimir effect," according to Margaret Hawton.[9] When I spoke with Dr. Harold Puthoff, from the Institute for Advanced Studies in Austin, Texas, he described his plans, along with Dr. Eric Davis, for measuring the electron levels of a hydrogen atom while it is in a Casimir cavity (figure 12.2). Puthoff proposed years ago that ZPE determines the ground state of hydrogen in the journal *Physical Review*. Recently, he arranged access to the facilities for testing his theory in a Casimir cavity that restricts the range of frequencies of ZPE around the hydrogen atoms.[10] According to Puthoff, this should show depressed or lowered electron levels as a result of less vacuum fluctuation activity of virtual particles that normally supports each electron level, in the same way that restricted virtual particle density inside a Casimir cavity causes the Casimir force to push inward.

A classic *New Scientist* journal article, with the same article title as this section, describes this phenomenon in more glorious detail. Zero point energy stabilizes the electron in a set ground-state orbit. It seems that the very stability of matter itself appears to depend on an underlying sea of electromagnetic zero point energy.[11]

LIMITATIONS ON
ZERO POINT ENERGY

While the facts support an unlimited energy plenum for the quantum vacuum, the arguments still support conservative, "right-wing physics." The actual estimates of energy density of even the limited ZPF (bounded by a maximum frequency) however are very impressive. For example, if we presume that the minimum possible wavelength is limited to the size of the proton, Nobel Prize–winning physicist Dr. Richard Feynman calculated that the energy density of the ZPF would be ten raised to the 108th power joules per cubic centimeter (10^{108} J/cc)!

Today, physicists want to look at even smaller vibration units like subatomic particles and so on. This makes the ZPF energy density escalate even more. In comparison, if we convert energy to mass using $E = mc^2$, we find that the equivalent "mass density" of the ZPF is ten to the ninety-fourth power grams per cubic centimeter (10^{94} g/cc). Compare that with typical nuclear densities of ten to the fourteenth power grams per cubic centimeter (10^{14} g/cc).[12] Therefore, gram for gram, ZPE offers almost ten to the eightieth times more energy for the same amount of space as nuclear reactors. Assuming they have the same efficiency, one ZPE engine equals 10^{80} nukes. It can be concluded that empty space contains more energy than matter does, for the same volume, even if we convert all of the matter into energy.

In a Los Alamos Laboratory, the Casimir force from ZPE was measured (in 1997) for the first time with conductive plates and found to be within 5 percent agreement with theory.[13] At AT&T Bell Labs, Federico Capasso designed a special semiconductor device that will only allow electrons to resonantly tunnel to the other side of a semiconductor layer because of the presence of zero point energy.[14] Zero point energy has been offered as a power source for the advanced energy and propulsion systems the government supposedly has in classified programs, to "produce vast amounts of energy without any pollution."[15]

New Scientist reported on NASA's Breakthrough Propulsion Physics First Project award to Illinois Professor Jordan Maclay "to study the energy of a vacuum."[16] In Europe, the University of Leicester, in the Virtual Microscopy Centre and the Nanoscale Interfaces Centre, has put the university in a key position to take a lead in Casimir force measurements in novel geometries. It led to the 2005 award of an €800,000 grant (NANOCASE) from the European framework 6 NEST (New and Emerging Science and Technology) program to lead a consortium from three countries (United Kingdom, France, and Sweden). The program will use the ultra-high-vacuum atomic force microscope installed in the Physics and Astronomy Department under Dr. Chris Binns to make very-high-precision Casimir force measurements in nonsimple cavities and assess the utility of the force in providing a method for contactless transmission in nanomachines.[17] The NANOCASE grant is the largest ZPE grant issued in the world so far.

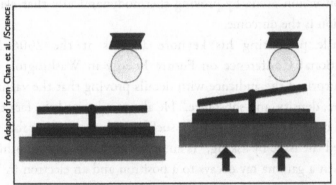

A chip with a see-saw plate suspended parallel to its surface (left) is pushed up (right) toward a ball. Quantum fluctuations in empty space produce a force that tilts the plate.

Figure 12.3. Nanoscale seesaw powered by ZPE.
Source: Science *magazine.*

THE CONVERSION OF ZPE

Now to explore the theme of this article, it is argued that the practical methods for converting ZPE to electricity are finally emerging. Localized or distributed electrical energy generation, as a result, looks much more promising than ever before. In 2001, *Popular Mechanics* featured an article talking about putting "free energy to work" while moving a nanoscale seesaw (figure 12.3), which activates the Casimir force at distances less than one micron (1,000 nm).[18] In 1998, a physicist formerly with the Jet Propulsion Laboratory, Fabrizio Pinto (who also worked with Robert Forward), invented and patented a nanoscale ZPE engine that obeys the laws of thermodynamics and pumps electrons, with the help of a tiny microlaser, just like an electrical generator. The effect of the microlaser pulses, which momentarily radically alters the Casimir force in the engine cavity, far exceeds the power input.

In the event of no other alternative explanations, one should conclude that major technological advances in the area of endless, by-product-free, free-energy production could be achieved.[19]

Pinto's accomplishment has brought much-needed legitimacy to the ZPE conversion arena by proving thermodynamically that electricity generation is the outcome.

While presenting his keynote address at the 2006 Second International Conference on Future Energy in Washington D.C.,[20] Pinto surprised the audience with details proving that the vacuum has "pressure, density, and substance." He also emphasized the fact that the vacuum can move physical objects, such as with the Casimir force. Pinto provoked the issue by asking, "Is empty space really worth nothing?"[21]

When a gamma ray decays to a positron and an electron in Wilson

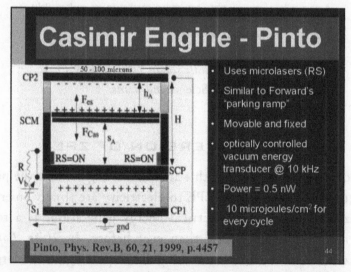

Figure 12.4. Casimir engine

cloud chamber experiments, the tracks left by the particle and antiparticle are clearly seen as they move away from each other with oppositely spinning paths.[22] Yet, the great quantum physicist Paul Dirac explains that in those days, people were very reluctant to postulate the existence of a new particle, even though they had seen it in the lab.

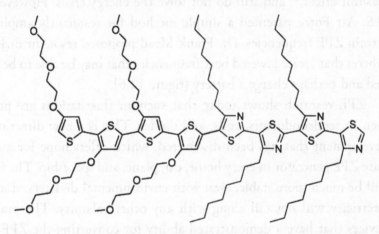

Figure 12.5. Electrically rectifying diodes come in nano-size (thin film) and pico-size (molecular).

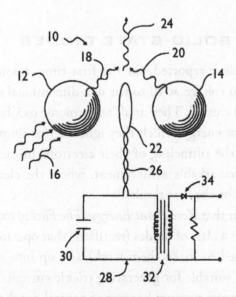

Figure 12.6. Drawing from Air Force ZPE Patent 5,590,031 by Dr. Frank Mead

When asked, "Why had experimenters never observed these anti-particles in cloud chambers?" Dirac's response was that a bias existed against new particles. This is very much like the present situation with ZPE. Physicists acknowledge the theory, publish papers on the subject, with an exponentially increasing number of papers devoted to the Casimir effect,[23] and still do not solve the energy crisis. However, the U.S. Air Force patented a simple method for resonantly amplifying certain ZPE frequencies. Dr. Frank Mead proposed resonant dielectric spheres that create lowered beat frequencies that may be able to be rectified and perhaps charge a battery (figure 12.6).

ZPE research shows today that vacuum fluctuations are present even in semiconductors, coils, and diodes. This is a new direction for development that has been discovered, which offers hope for a solid-state ZPE generator in every home, car, plane, and spaceship. The future will be much more stable, even with environmental disasters, since the electricity will not fail along with any other calamity. The practical devices that have a demonstrated ability for converting the ZPE non-thermal noise into useful electrical generation use semiconductors, coils, or diodes, with diodes being the most convenient.

SOLID-STATE DIODES

In 1994, J. Smoliner reported, for the first time, resonant tunneling while applying no voltage at all to the one-dimensional quantum wells that his team had created. They used "anharmonic oscillation" to substitute for zero point energy, which they ignored "for simplicity," though it was powering the tunneling of their electrons in each well. Figure 12.7 shows the remarkable achievement, where the electrons prefer a zero voltage bias for the best results.[24]

The argument that *Zero Point Energy: The Fuel of the Future* makes is that there exists a class of diodes (rectifiers) that operate at "zero bias" (no voltage applied to make them work) and up into microwave frequencies that are suitable for generating trickle currents from the zero point energy quantum vacuum because of natural nonthermal electrical fluctuations (Johnson noise).

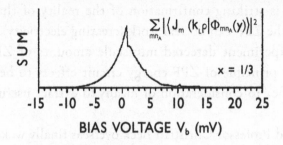

$$\sum_{mn_A} \left| \langle J_m \left(k_{LP} | \Phi_{mn_A}(y) \rangle \right| ^2$$

x = 1/3

BIAS VOLTAGE V_b (mV)

Figure 12.7. Chart from a German research project showing an electron tunneling through a barrier at zero bias voltage

Furthermore, there are peer-reviewed journal articles that also show tunneling at zero voltage (zero bias). Several microwave diodes in the book excerpt also exhibit this feature. However, you have to appreciate that the noise level (1/f noise or Johnson noise) is where ZPE manifests. (See *Practical Conversion of Zero-Point Energy: Feasibility Study of the Extraction of Zero-Point Energy from the Quantum Vacuum for the Performance of Useful Work* for a more rigorous treatment).[25] Nature has also been helpful since Johnson noise in the diode is also generated at the junction itself and therefore requires no minimum signal to initiate the conduction in one direction.

SUBSTANTIVE EXPERIMENTS

The following U.S. patents are the most significant in ZPE research: Rectifying thermal electric noise by Charles Brown, Patent 3,890,161, and Patent 4,704,622 by Capasso, which actually acknowledge ZPE for their functional nature. Metal-insulator-metal nanodiodes probably will be a popular brand for ZPE usage with millipore sheet assembly, as Brown suggests.

These diodes demonstrate substantive, greater-than-uncertainty generation of energy from ZPE. In fact, simple coils do as well, according to the published articles by Roger H. Koch and colleagues.[26] In a series of experiments from 1980 to 1982, Koch measured voltage fluctuations in resistive wire circuits that are induced by the ZPF. The

Koch result is striking confirmation of the reality of the ZPF and proves that the ZPF can do real work (creating electricity). Although the Koch experiment detected miniscule amounts of ZPF energy, it shows the principle of ZPF energy circuit effects to be valid and opens the door to consideration of means to extract useful amounts of energy.

Davis and Professor Christian Beck overseas finally woke up to the multiple papers that Koch published. Davis announced his efforts to obtain Lockheed money to fund a replication of Koch's work.[27] Beck wrote a book on ZPE after he published a paper about dark energy being measurable in the laboratory.[28]

Pinto in 1999 and others like the Brown patent make reasonable calculations of the predicted energy density of an array of vacuum engines and ZPE diodes, which are about one kilowatt per square meter and conservatively reach estimates of hundreds of kilowatts per cubic meter for stacked arrays.

CUSTOM-MADE ZERO BIAS DIODES

Other diodes that exhibit the ability to rectify electromotive force (EMF) energy include the class of "backward diodes" that operate with zero bias (no external power supply input). (See U.S. Patent 6,635,907, Type II interband heterostructure backward diodes, and also U.S. Patent 6,870,417, Circuit for loss-less diode equivalent.) These have been used in microwave detection for decades and have never been tested for nonthermal zero point energy fluctuation conversion. There is every reason to presume they include such ZPE radiation conversion in their everyday operation but it is unnoticed with other EMF energy, being so much larger in amplitude. U.S. Patent 6,635,907 from HRL Laboratories describes a diode with a very desirable, "highly nonlinear portion of the I-V curve near zero bias." These diodes produce a significant current of electrons when microwaves in the gigahertz range are present. Another example is U.S. Patent 5,930,133 from Toshiba, titled Rectifying device for achieving high power efficiency. They use a tunnel diode in the backward mode so that "the turn-on voltage is zero." Could there be a better device for small-voltage ZPE fluctuations that don't like to jump big barriers?

A completely passive, unamplified zero bias diode converter/detector for millimeter (GHz) waves was developed by HRL Laboratories in 2006 under a Defense Advanced Research Projects Agency contract, using an Sb-based "backward tunnel diode." It is reported to be a "true zero-bias diode" that does not have significant 1/f noise when it is unamplified. It was developed for a "field radiometer" to "collect thermally radiated power" (in other words, "night vision"). The diode array mounting allows a feed from a horn antenna, which functions as a passive concentrating amplifier. The important clue is the "noise equivalent power" of 1.1 pW per root hertz (picowatts are a trillionth of a watt) and the "noise equivalent temperature difference" of 10K, which indicate a sensitivity to Johnson noise, the source of which is ZPE. Perhaps HRL Laboratories will consider adapting the invention for passive zero point energy generation.[29]

Another invention developed in 2005 by the University of California, Santa Barbara, is the "semimetal-semiconductor rectifier" for similar applications, to rival the metal-semiconductor (Schottky) diodes that are more commonly known for microwave detection. These zero bias diodes can operate at room temperature and have a noise equivalent power of about 0.1 pW but a high "RF-to-DC current responsivity" of about 8 A/W (amperes per watt). Most importantly, the inventors claim that the new diodes are about 20 dB more sensitive than the best available zero bias diodes from Hewlett-Packard.[30]

There also have been other inventions such as "single electron transistors" that also have "the highest signal-to-noise ratio" near zero bias. Furthermore, "ultrasensitive" devices that convert radio frequencies have been invented that operate at outer space temperatures (three degrees above zero point: 3°K). These devices are tiny nanotech devices, so it is possible that lots of them could be assembled in parallel (such as an array) to produce ZPE electricity with significant power density.[31]

MOLECULAR REFRIGERATORS

Nonthermal ZPE is known to work even at temperatures near zero degrees Kelvin, affecting every particle kinetically in the same manner as thermal energy. Liquid helium stays liquid at microdegrees above absolute zero for this reason. Furthermore, it is well understood that a proposed design for a zero bias diode rectifier circuit that generates electricity from ZPE will also rectify and convert thermal noise as well. In other words, heat is absorbed and also adds power to these proposed ZPE converters, which cools the space around them.

Environmentally, in this age of global warming, what better device to have in every home and business than a solid-state, self-sustaining refrigeration unit that generates electricity? That is exactly what has been independently discovered in the molecular arena by Professor Chris Van den Broeck from Hasselt University in Belgium.[32] Figure 12.8 is from his *Physical Review Letters* article that shows how such molecules will work. He describes them as a paddlewheel type of model (figure 12.8a), which the real molecular machine simulates (figure 12.8b). The exciting prediction is that a small thermal gradient will cause them to work or a mechanical force (0.1 piconewtons) will tend to create refrigeration. Thus, molecular refrigerators and solid-state ZPE diodes work in the same manner.

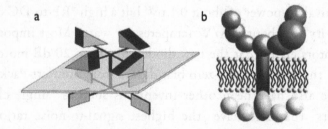

a   b

*Figure 12.8. Molecular refrigerators that transfer heat
to rotational energy will also convert ZPE.*

MOVEMENT FROM NOTHING

Empty space can set objects in motion, a physicist claims.[33] Motion can be conjured out of thin air, according to a physicist in Israel. Alexander

Feigel of the Weizmann Institute of Science in Rehovot says that objects can achieve speeds of several centimeters an hour by getting a push from the empty space of a vacuum.[34]

His theory and experiment offer a possible explanation for the accelerated expansion of distant galaxies, as well as the motion of fluids.[35] Furthermore, Feigel's discovery opens the way to harness virtual particles to do useful work by creating propulsion.

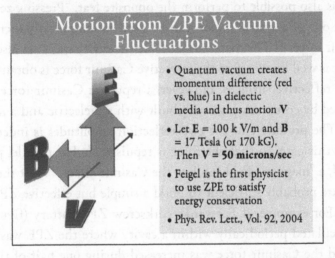

Figure 12.9. My lecture slide explaining the Feigel discovery that ZPE conservation of energy can include a propulsive force in the presence of a static magnetic and electric field

ADDITIONAL ZPE CONVERSION METHODS

Another interesting experiment is the Casimir effect at macroscopic distances, which proposes observing the Casimir force at a distance of a few centimeters using confocal optical resonators within the sensitivity of laboratory instruments.[36] This experiment makes the microscopic Casimir effect observable and greatly enhanced. Then, any other modes of conversion for the Casimir force should very well be improved.

In general, many of the experimental journal articles refer to vacuum effects on a cavity that is created with two or more surfaces. Cavity quantum electrodynamics (QED) is a science unto itself. "Small cavities suppress atomic transitions; slightly larger ones, however, can enhance them. When the size of the cavity surrounding an excited atom is increased to the point where it matches the wavelength of the photon that the atom would naturally emit, vacuum-field fluctuations at that wavelength flood the cavity and become stronger than they would be in free space."[37]

It is also possible to perform the opposite feat. "Pressing zero-point energy out of a spatial region can be used to temporarily increase the Casimir force."[38] The materials used for the cavity walls are also important. It is well known that the attractive Casimir force is obtained from highly reflective surfaces. However, a repulsive Casimir force may be obtained by considering a cavity built with a dielectric and a magnetic plate. The product r of the two reflection amplitudes is indeed negative in this case, so that the force is repulsive."[39] For parallel plates in general, a "magnetic field inhibits the Casimir effect."[40] Just these tools alone are probably sufficient to build a simple but effective ZPE transducer. For example, if Forward's corkscrew ZPE battery (figure 12.1) was oscillated periodically within a cavity where the ZPE was pressed out and the Casimir force was increased during one half of the cycle, the spring would tend to attract and force electrons out of the battery by electrostatic repulsion. Then, during the second half of the cycle, electrons would be drawn back into the battery as the spring relaxed, with the Casimir force decreasing. A simple diode rectifier with zero bias voltage in the external circuit will rectify this alternating current (AC) and produce usable electricity. Of course, efficiency calculations and measurements are needed to ensure the overunity production of energy that costs less to produce than the energy generated.

Forward, who passed away in 2002, said:

> Before I wrote the paper everyone said that it was impossible to extract energy from the vacuum. After I wrote the paper, everyone had to acknowledge that you could extract energy from the vacuum, but began to quibble about the details. The spiral design won't work very efficiently. . . . The amount of energy extracted is extremely

small. . . . You are really getting the energy from the surface energy
of the aluminum, not the vacuum. . . . Even if it worked perfectly,
it would be no better per pound than a regular battery. . . . Energy
extraction from the vacuum is a conservative process, you have to
put as much energy into making the leaves of aluminum as you will
ever get out of the battery. . . . Yes, it is very likely that the vacuum
field is a conservative one, like gravity. But, no one has proved it yet.
In fact, there is an experiment mentioned in my Mass Modification
paper (military report on a proposed experiment for an antiproton
in a vacuum chamber), which can check on that.

The amount of energy you can get out of my aluminum foil battery is limited to the total surface energy of all the foils. For foils
that one can think of making that are thick enough to reflect ultraviolet light so the Casimir attraction effect works, say 20-nm (70
atoms) thick, then the maximum amount of energy you get out
per pound of aluminum is considerably less than that of a battery.
To get up to chemical energies, you will have to accrete individual
atoms using the van der Waals force, which is the Casimir force for
single atoms instead of conducting plates. My advice is to accept the
fact that the vacuum field is probably conservative and invent the
vacuum equivalent of the hydroturbine generator in a dam. [41]

To me this was very interesting advice, for someone like me just starting out in vacuum engineering in the 1990s. Forward is probably the first
physicist to give considerable thought to the subject of this book, so he
deserves some recognition for being a pioneer of ZPE applications. As it
turns out, the spiral design is more of a battery than a generator. Also,
the mounting evidence for the ZPF is weighing heavily in favor of being
a nonconservative field. This means the experts are finding more ways to
extract energy from ZPE without putting energy in first.

CASIMIR ENGINE

The Casimir force presents a fascinating exhibition of the power of the
ZPF, offering about one atmosphere of pressure when plates are less

than one micron apart. As is the case with magnetism today, it has not been immediately obvious, until recently, how a directed Casimir force might be cyclically controlled to do work. The optically controlled vacuum energy transducer of figure 12.4, however, proposed by Pinto presents a powerful theoretical invention for rapidly changing the Casimir force by a quantum surface effect, excited by photons, to complete an engine cycle and thus transfer a few electrons. The convincing part of Pinto's invention is the quantum mechanics and thermodynamics that he brings to the analysis, offering a conclusive engine for free energy production. Pinto's company, Interstellar Technologies Corporation (www.Interstellartechcorp.com), is dedicated to "turning quantum vacuum engineering into a commercially viable activity."[42]

Figure 12.10. Microdisk one micron in diameter

Pinto uses mechanical forces from the Casimir effect and a change of the surface dielectric properties to intimately control the abundance of virtual particles. Basically, it is an optically controlled vacuum energy transducer. A moving cantilever or membrane is proposed to cyclically change the active volume of the chamber as it generates electricity with a thermodynamic engine cycle. The invention proposes to use the Casimir force to power the microcantilever beam produced with standard micromachining technology. The silicon structure may also include a microbridge or micromembrane instead, all of which have a natural oscillation frequency on the order of a free-carrier lifetime in the same material.

The invention is based on the cyclic manipulation of the dimensions of the Casimir cavity created between the cantilever and the substrate. The semiconducting membrane is the cantilever, which could be on the order of 50 to 100 microns in size, with a few-micron thickness in order to obtain a resonant frequency in the range of 10 kHz, for example.

Two monochromatic lasers are turned on, thereby increasing the Casimir force by optically changing the dielectric properties of the cantilever. There is a frequency dependence of a dielectric constant that can vary with frequency by a few orders of magnitude, but inversely proportional to the frequency. This means that as the frequency goes up, the dielectric constant may go down.

Pinto's proactive approach is to excite a particular frequency mode in the cavity. In doing so, an applied electrostatic charge increases as the cantilever is pulled toward the adjacent substrate by the Casimir force. Bending the charged cantilever on a nanoscale, the Casimir attractive force is theoretically balanced with opposing electrostatic forces, in the same way as Forward's "parking ramp," seen in other chapters.

Analysis of the Casimir engine cycle (see *ZeroPoint Energy: Fuel of the Future* for analysis) demonstrates its departure from hydroelectric, gaseous, or gravitational systems. For example, the Casimir pressure always acts opposite to the gas pressure of classical thermodynamics and the energy transfer that causes dielectric surface changes "does not flow to the virtual photon gas." Altering physical parameters of the device, therefore, can change the total work done by the Casimir force, in contrast to gravitational or hydroelectric systems. Unique to the quantum world, the type of surface and its variation with optical irradiation is a key to the transducer operation. Normally, changing the reflectivity of a surface will affect the radiation pressure on the surface but not the energy density of the real photons. However, in the Casimir force case, Pinto explains, "The normalized energy density of the radiation field of virtual photons is drastically affected by the dielectric properties of all media involved via the source-free Maxwell equations."

Specifically, Pinto discovered that the absolute value of the vacuum energy can change "just by causing energy to flow from a location to another inside the volume V." This finding predicts a major breakthrough in use of a quantum principle to create a transducer of vacuum energy. Some concerns are usually raised, as mentioned previously, with whether the vacuum energy is conserved. In quantum systems, if the parameters (boundary conditions) are held constant, the Casimir force is strictly conservative in the classical sense, according to Pinto. "When they are changed, however, it is possible to identify closed paths along which the total work done by this force does not vanish."

To conclude the energy production analysis, it is noted by Pinto that 10,000 cycles per second are taken as a performance limit. Taking the lower estimate of 100 erg/cm² per cycle, power or "wattage" is calculated to be about 1 kW/m², which is about the same as photovoltaic energy production. However, this invention will work twenty-four hours a day, seven days a week and is not dependent on the sun. The single cantilever transducer is expected to produce about 0.5 nW and establish a millivolt across a kilo-ohm load, which is still fairly robust for such a tiny machine.

The basis of the dielectric formula starts with Pinto's analysis that the Drude model of electrical conductivity is dependent on the mean electron energy and estimated to be in the range of submillimeter wavelengths. The Drude model, though classical in nature, is often used for comparison purposes in Casimir calculations. The detailed analysis by Pinto shows that carrier concentrations and resistivity contribute to the estimate of the total dielectric permittivity function value, which is frequency dependent. The frequency dependence is of increasing concern for investigations into the Casimir effects on dielectrics. A higher-frequency laser may have significantly different effects than an infrared laser, for example.

Another suggested improvement to the original invention could involve a femtosecond or attosecond pulse from a disk-shaped semiconductor microlaser, such as those developed by Bell Laboratories (figure 12.10). The microlaser could be used in close proximity to the cantilever assembly. Such microlaser structures, called microdisk lasers, measuring 2 microns across and 100 nm thick, have been shown to produce coherent light radially. An optimum choice of laser frequency would be to tune it to the impurity ionization energy of the semiconductor cantilever. In this example, the size would be approximately correct for the micron-sized Casimir cavity, according to the research that I have done investigating this invention.

Pinto chooses to neglect any temperature effects on the dielectric permittivity. However, since then, the effect of finite temperature has been found to be intimately related to the cavity edge choices that can cause the Casimir energy to be positive or negative. Therefore, the contribution of temperature variance and optimization of the operating temperature seems to have become a parameter that should not be ignored. Also supporting this view is the evidence that the dielectric permittivity has been found to also depend on the derivative of the dielectric permittivity with respect to temperature.

The microfabrication task for the Casimir engine includes mounting microlasers inside the Casimir cavity and ensuring that an extended or continuous 10-KHz repetition rate is possible with a moving cantilever by addressing the expected lifespan. It is worth mentioning in support of Pinto that similar cantilevers, made from silicon of the same micron size, with only one support end, now operate in many new automobiles as acceleration and crash sensors, without high failure rates. The energy production rate for the Casimir engine is predicted to be fairly robust (0.5 nW per cell or 1 kW/m^2), which could motivate a dedicated research and development project in the future. However, the Casimir engine project of Pinto's appears to be a million-dollar investment at best, which is to be expected for something so revolutionary. Pinto's business plan probably addresses the full-scale production costs and projected payback.

Using some of the latest cavity QED techniques, such as mirrors, resonant frequencies of the cavity vs. the gas molecules, quantum coherence, vibrating cavity photon emission, rapid change of refractive index, spatial squeezing, cantilever deflection enhancement by stress, and optimized Casimir cavity geometry design, the Pinto invention may be improved substantially. The process of laser irradiation of the cavity, for example, could easily be replaced with one of the above-mentioned quantum techniques for achieving the same variable dielectric and Casimir force effect, with less hardware involved. At the present stage of theoretical development, the Pinto device receives a moderate rating of feasibility. Its overall energy quality rating, in my opinion, is very high.

WHAT LIES BENEATH THE VOID— NANOCASE PROJECT

A cursory summary of some of the physics behind the ZPE effects is now presented for the nontechnical audience. A more detailed, rigorous review is to be found in my feasibility study called *Practical Conversion of Zero-Point Energy* (available from Amazon). Let's begin with a look at a British development called NANOCASE that supports ZPE research on a very significant budget scale. It simply looks for better measurements of the Casimir force (see next section, "Casimir Force"), which is a start in the right direction. Professor Chris Binns (physics and astronomy) at the University of Leicester received European funding in 2005 for an exciting project to measure the force of zero point energy.

The e-bulletin story from the United Kingdom on NANOCASE

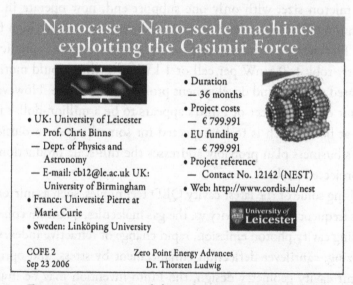

Nanocase - Nano-scale machines exploiting the Casimir Force

- UK: University of Leicester
  — Prof. Chris Binns
  — Dept. of Physics and Astronomy
  — E-mail: cb12@le.ac.uk UK: University of Birmingham
- France: Université Pierre at Marie Curie
- Sweden: Linköping University

- Duration
  — 36 months
- Project costs
  — € 799.991
- EU funding
  — € 799.991
- Project reference
  — Contact No. 12142 (NEST)
- Web: http://www.cordis.lu/nest

University of Leicester

COFE 2   Zero Point Energy Advances
Sep 23 2006   Dr. Thorsten Ludwig

Figure 12.11. Lecture slide explaining the million-dollar grant for zero point energy research

and Binn has the subtitle "Exciting Project Connected to the 'Zero-Point Energy' of Space." While NANOCASE is the largest ZPE grant so far, the details of the Casimir force, including its positive-to-negative force reversal, have already been worked out by Dr. Jordan Maclay with a NASA grant years ago (discussed in detail in the book *Zero Point Energy: The Fuel of the Future*). It is true that the replication and verification of previously theorized results will advance this science, even if only by a snail's pace.

CASIMIR FORCE

It is only appropriate and fitting that the actual simple-but-elegant equation for the Casimir force is also included (figure 12.12). It depends on three constants (Planck's constant, pi, and the speed of light, where h = h/2) and, of course, the distance L between the plates. There are very few relationships in nature that depend on the fourth power, but here we see one. This causes all of the problems of stiction, with such a dramatic increase in the force as the separation distance L gets really tiny. However, the gecko is happy since this effect lets him stick to any surface (even the Science Channel recently showed a live gecko in a lab and a close-up of the tiny hairs on its feet, which they related to van der Waals forces).

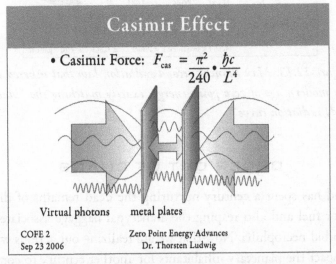

Figure 12.12. *The Casimir force equation producing attraction*

ZERO POINT ENERGY BASICS

Max Planck, a Nobel Prize winner, corrected his mistake by introducing a second radiation law that made all the difference in the world. Seen in all of its glory in figure 12.13, the important point for nonspecialists is the circled quantity ½ hf. This is the mysterious average of zero point energy when looked at with quantum mechanics. I say "mysterious" because it is exactly half of what physicists normally call a quantum. For example, Planck's quantum frequency equation, $E = hf$, gives a clear indication of the quantum units of hf that are present in an atom (with any multiples also allowed by the same equation). However, this is nature that determined the average, and it is also proven or derived from the Heisenberg uncertainty principle.

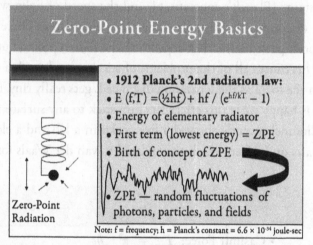

Zero-Point Energy Basics

- **1912 Planck's 2nd radiation law:**
- $E\,(f,T) = \boxed{½hf} + hf\,/\,(e^{hf/kT} - 1)$
- Energy of elementary radiator
- First term (lowest energy) = ZPE
- Birth of concept of ZPE
- ZPE — random fluctuations of photons, particles, and fields

Zero-Point Radiation

Note: f = frequency; h = Planck's constant = 6.6×10^{-34} joule-sec

Figure 12.13. Max Planck's second radiation law that ushered in the modern age of zero point energy, exactly matching the "black body radiation curve"

ORDER OUT OF CHAOS

Mankind has spent a century nurturing the dead remains of dinosaur fossils for fuel and also reaping the archetypal baggage associated with such morbid necrophilia. Now that we are realizing our actions will seriously impact the planetary inhabitants for another century to come, it is time to look to a more vital energy source. As an alternative, it can be argued that ZPE is a living energy in the broad sense of the word. The ZPF breathes life into every atom, sustaining its size and shape, and it certainly is renewable. Further insights can be gained from the chaotic activity that is at the core of the ZPF. Studying chaos theory, one learns about "strange attractors," "islands of stability," and "order out of chaos." In fact, the Nobel Prize winner Ilya Prigogine wrote a book with the title *Order Out of Chaos,* which is an excellent discourse on the topic of this section.[43] When faced with nonlinear chaotic systems like the ZPF, Prigogine makes the case that such systems will often find stability "far from equilibrium." With such a counter-intuitive notion, supported by many specific examples in nature, Prigogine explains the title of his book by saying,

"At all levels, be it the level of macroscopic physics, the level of fluctuations, or the microscopic level, nonequilibrium is the source of order. Nonequilibrium brings 'order out of chaos.'"[44]

He also delves into an excellent discussion of entropy and the second law of thermodynamics. Prigogine also explains irreversibility with a rediscovery of the fluctuation-dissipation theorem, without apparently knowing about H. B. Callen's earlier publication. "Intrinsic irreversibility is the strongest property: it implies randomness and instability."[45] Thus we have another fundamental description of ZPE from chaos theory, which also includes a description of the amplification of fluctuations as well.[46] Extracting energy from ZPE has just become easier, in so many ways, so that our future will be much brighter and offer much more freedom for everyone.

CONCLUSIONS

Recently, a single nanowire of silicon 300 nm wide has been proven by Charles Lieber from Harvard University to absorb light and create electricity. In response, Phaedon Avouris, a fellow at IBM Research, said, "There has been a lot of talk recently about making independent nanomachines and nanosystems. The issue has always been, how are you going to power them? If you want to have an independent nanosystem that's self-contained, that's not plugged into a central power supply, then you need something like this."[47] The same can be said for zero point energy converters, which surprisingly work best and with the most power density, as Dr. Frank Mead pointed out, when made very small.

GLOSSARY

Following are terms that are used throughout the article:

Casimir force: A force attributed to zero point energy fluctuations that is usually attractive and measured between two parallel plates or any two parallel surfaces that come closer than 1 micron apart (1 millionth of a meter). It is the force that keeps a gecko attached to a wall and most insects as well.

Diodes: The solid-state device that is able to send electrical current in one direction only (DC), even if the input is AC electricity, accomplishing "rectification" of the input.

Energy: The capacity for doing work. Equal to power exerted over time (e.g., kilowatt-hours). It can exist in linear or rotational form and is quantized in the ultimate part. It may be conserved or not conserved, depending on the system considered. Mostly, all terrestrial manifestations can be traced to solar origin, except for zero point energy.

Lamb shift: A shift (increase) in the energy levels of an atom, regarded as a Stark effect due to the presence of the zero point field. Its explanation marked the beginning of modern quantum electrodynamics.

Nanoscale: A small scale of measurement related to nanotechnology (10-9 meters) that is a thousand times smaller than a micron and a source of great innovation today in machines and materials (e.g., carbon nanotubes).

Planck's constant: The fundamental basis of quantum mechanics that provides the measure of a quantum ($h = 6.6 \times 10^{-34}$ joule-seconds); it is also the ratio of the energy to the frequency of a photon.

Quantum electrodynamics: The quantum theory of light as electromagnetic radiation, in wave and particle form, as it interacts with matter. Abbreviated "QED."

Quantum vacuum: A characterization of empty space by which physical particles are unmanifested or stored in negative energy states. Also called the physical vacuum.

Uncertainty principle: The rule or law that limits the precision of a pair of physical measurements in complementary fashion (e.g., the position and momentum, or the energy and time), forming the basis for zero point energy.

Virtual particles: Physically real particles emerging from the quantum vacuum for a short time, determined by the uncertainty principle. This can be a photon or other particle in an intermediate state that, in quantum mechanics (Heisenberg notation), appears in matrix elements connecting initial and final states. Energy is not conserved in the transition to or from the intermediate state. Also known as a virtual quantum.

Zero point energy: The nonthermal, ubiquitous kinetic energy (averaging ½ hf) that is manifested even at zero degrees Kelvin, abbreviated as "ZPE." Also called vacuum fluctuations, zero point vibration, residual energy, quantum oscillations, the vacuum electromagnetic field, virtual particle flux, and, recently, dark energy.

Zitterbewegung: An oscillatory motion of an electron, exhibited mainly when it penetrates a voltage potential, with frequency greater than 1,021 Hz. It can be associated with pair production (electron-positron) when the energy of the potential exceeds $2 mc^2$ (m = electron mass). Also generalized to represent the rapid oscillations associated with zero point energy.

AFTERWORD

Brian O'Leary, Ph.D.

EDITOR'S NOTE: Many of us have profoundly felt the loss of Brian O'Leary and his passionate voice on behalf of new energy technologies and the role they can play in radically transforming our planet, halting environmental pollution, and allowing the rise of poor nations who, without access to free energy, have little hope of escaping poverty. Although Brian and I never met, I've missed our many e-mail exchanges on a wide range of shared interests. It seems appropriate to end *Infinite Energy Technologies* with Brian's wise counsel. May we all heed his words. —Finley Eversole, Ph.D.

SUPPORT IS NEEDED— NOT DENIAL AND SUPPRESSION

Say you felt that practical free energy might be possible—and you don't have to be a believer at the outset—would you support the R and D, would you ignore it, or would you want to suppress it in order to protect your own current interests? Are you ready to let go of the past and join the dedicated evolutionary energy team? Are you ready to become a hero who might risk wealth and reputation? The choice is yours. And make no mistake, you will choose either by your action or inaction. This is your moment of choice.

I am happy to see Mongolia and other nations committing to socially and environmentally sustainable programs like the Exemplar

Zero Initiative. The true purpose of governments should be to protect and enhance the lives of their people and ecosystems, to support *buen vivir* (good living) for generations to come.

The global economic and political landscape is shifting rapidly into a more multipolar world as the one remaining superpower, the United States, declines under its own economic weight and its overextended military hegemony.

There is good news and bad news about this development. On the one hand, nations, regions, and municipalities now have the opportunity to become more independent from international corporate and governmental pressures through sustainable innovation. On the other hand, the temptation to allow multinational corporations to exploit nonrenewable resources for export becomes ever greater as developing countries seek short-term economic gain so they can service debts and enhance infrastructure and social programs.[1]

This approach is a losing proposition in the long run. What happens when our resources and ecosystems are depleted in the wake of such shortsighted thinking? Unfortunately, most corporations and governments measure their performance based on quarterly profits and the terms of elective political office and not on long-term time horizons like the Native Americans' planning seven generations ahead.

A prime example of this dynamic is Ecuador, where I live. The story of Ecuador is the familiar story of numerous countries that have depended on the temporary export of oil, gas, minerals, and lumber to sustain their economies. Chevron-Texaco is responsible for leaving behind massive and devastating amounts of toxic waste in the Ecuadorian Amazon, resulting in what is the largest environmental lawsuit in world history, with claims of $27 billion in damages for the diseases and wrongful deaths of thousands of people living in the area.[2] The threat to the environment posed by imminent oil drilling in the western Amazon region is extremely serious and widespread, with over one hundred leased "oil blocks" embracing much of the rain forests of Ecuador, Peru, Colombia, Brazil, and Bolivia.[3]

To its credit, the government of Ecuador proposed keeping the oil in the ground in one oil block in the biodiverse rain forest of

Yasuni National Park if the international community were to match funds to cover potential lost revenues.[4] That's a good start, but it isn't enough. Eighty percent of the Ecuadorian and Peruvian Amazon is earmarked for oil and gas drilling, with associated road building and deforestation.

Currently, one-third of Ecuador's total revenues come from petroleum exports. Some of us are proposing that these revenues be replaced by income from new energy technologies, sustainable agriculture, medicinal herbs from the rain forest, innovative water treatment, ecotourism, health tourism, and the acquisition of conservation land trusts through carbon credits and gifts.[5]

We firmly believe that an open acceptance of innovation has the potential to generate sufficient income so that Ecuador can leave the rain forest and its indigenous peoples alone while creating economic sovereignty for itself. We propose that innovation sanctuaries be established, protected by the government, to allow researchers and entrepreneurs to do the necessary R and D on their technologies, the most promising of which would be implemented.

The biggest global challenge we face is to cocreate from an altruistic perspective those social, political, and economic systems to foster the needed systemic changes in our governments and corporations. So far, we've been falling way short of the mark in all respects.

But we're going to need resources to begin the task.

This is an invitation to those of you who feel the call to redirect your abundant resources accumulated from an oil economy to alternative energy and other solutions that are truly clean and sustainable, to put your petrodollars to work, and to make your great-grandchildren proud of you. As Nelson Mandela said: "Sometimes it falls upon a generation to be great."

Ours is the generation that *must* be great.

Make no mistake: we're in the midst of an evolution of consciousness. We can cocreate a planet that works for everyone. We can redirect success from profit and pollution to true sustainability. We can reactivate an eden in which our new fertile crescents and restored ecosystems grow and grow while we give back to the Earth rather than take from it.

Technology can provide elegant answers to our desperate quandary, but the Taker culture has effectively blocked these solutions because of the greed of the few and the ignorance of the many, and so we find ourselves on a sinking ship. We've been denying the best possibilities because of economic self-interest, the fear of the unknown, and a fundamental reluctance to embrace bold new possibilities that await us if only we have a closer look.

Will we collectively shift the paradigm by introducing clean breakthrough energy, pure and abundant water, and the best quality of life for all of us? It's up to each of you to choose to do so fearlessly with love and compassion for all creation. You are the answer to the question. You are the points of conscious evolution.

I'd like to leave you with a quotation from what you can gather is one of my true heroes, Buckminster Fuller: "If the success or failure of this planet, and of human beings, depended on how I am and what I do, how would I be? What would I do?"

APPENDIX A

THE FOUR
OCCUPATIONS OF
PLANET EARTH

Tom Engelhardt

On the streets of Moscow in the tens of thousands, the protesters chanted, "We exist!" Taking into account the comments of statesmen, scientists, politicians, military officials, bankers, artists, all the important and attended-to figures on this planet, nothing caught the year more strikingly than those two words shouted by massed Russian demonstrators.

"We exist!" Think of it as a simple statement of fact, an implicit demand to be taken seriously (or else), and undoubtedly an expression of wonder, verging on a question: "We exist?"

And who could blame them for shouting it? Or for the wonder? How miraculous it was. Yet another country long immersed in a kind of popular silence suddenly finds voice, and the demonstrators promptly declare themselves not about to leave the stage when the day—and the demonstration—ends. Who guessed beforehand that perhaps fifty thousand Muscovites would turn out to protest a rigged electoral process in a suddenly restive country, along with crowds in St. Petersburg, Tomsk, and elsewhere from the south to Siberia?

In Tahrir Square in Cairo, they swore, "This time we're here to stay!" Everywhere this year, it seemed that they—"we"—were here to stay. In New York City, when forced out of Zuccotti Park by the police, protesters returned carrying signs that said, "You cannot evict an idea whose time has come."

And so it seems, globally speaking. Tunis, Cairo, Madrid, Madison, New York, Santiago, Homs. So many cities, towns, places. London, Sana'a, Athens, Oakland, Berlin, Rabat, Boston, Vancouver . . . it could take your breath away. And as for the places that aren't yet bubbling—Japan, China, and elsewhere— watch out in 2012 because, let's face it, "we exist."

Everywhere, the "we" couldn't be broader, often remarkably, even strategically, more ill defined; 99 percent of humanity contains so many potentially conflicting strains of thought and being: liberals and fundamentalists, left-wing radicals and right-wing nationalists, the middle class and the dismally poor, pensioners and high-school students. But the "we" couldn't be more real.

This "we" is something that hasn't been seen on this planet for a long time, and perhaps never quite so globally. And here's what should take your breath away, and that of the other 1 percent, too: "we" were never supposed to exist. Everyone, even "we," counted us out.

Until last December, when a young Tunisian vegetable vendor set himself alight to protest his own humiliation, that "we" seemed to consist of the nonactors of the twenty-first century and much of the previous one as well. We're talking about all those shunted aside, whose lives only weeks, months, or, at most, a year ago, simply didn't matter—all those the powerful absolutely knew they could ride roughshod over as they solidified their control of the planet's wealth, resources, property, as, in fact, they drove this planet down.

For them, "we" was just a mass of subprime humanity that hardly existed. So of all the statements of 2011, the simplest of them—"We exist!"—has been by far the most powerful.

NAME OF THE YEAR:
OCCUPY WALL STREET

Every year since 1927, when it chose Charles Lindbergh for his famed flight across the Atlantic, *Time* magazine has picked a "man" (even when, on rare occasions, it was a woman like Queen Elizabeth II) or, after 1999, a "person" of the year (though sometimes it's been an inanimate object like "the computer" or a group or an idea). If you want a gauge of how "we" have changed the global conversation in just months, those in the running this year included "Arab youth protestors," "Anonymous," "the 99 percent," and "the 1 percent." Admittedly, so were Kim Kardashian, Casey Anthony, Michele Bachmann, Kate Middleton, and Rupert Murdoch. In the end, the magazine's winner of 2011 was "the protester."

How could it have been otherwise? We exist—and even *Time* knows it. From Tunis in January to Moscow in December this has been, day by day, week by week, month by month, the year of the protester. Those looking back may see clues to what was to come in isolated eruptions like the suppressed green movement in Iran or under-the-radar civic activism emerging in Russia. Nonetheless, protest, when it arrived, seemed to come out of the blue. Unpredicted and unprepared for, the young (followed by the middle-aged and the old) took to the streets of cities around the globe and simply refused to go home, even when the police arrived, even when the thugs arrived, even when the army arrived, even when the pepper spraying, the arrests, the wounds, the deaths began and didn't stop.

And by the way, if "we exist" is the signature statement of 2011, the name of the year would have to be Occupy Wall Street. Forget the fact that the place occupied, Zuccotti Park, wasn't on Wall Street but two blocks away and that, compared to Tahrir Square or Moscow's thoroughfares, it was one of the smallest plots of protest land on the planet. It didn't matter.

The phrase was blowback of the first order. It was payback, too. Those three words instantly turned the history of the last two decades upside down and helped establish the protesters of 2011 as the third of the four great planetary occupations of our era.

Previously, "occupations" had been relatively local affairs. You occupied a country ("the occupation of Japan"), usually a defeated or conquered one. But in our own time, if it were left to me, I'd tell the history of humanity, American-style, as the story of four occupations, each global in nature.

THE FIRST OCCUPATION

In the 1990s, the financial types of our world set out to "occupy the wealth," planetarily speaking. These were, of course, the globalists, now better known as the neoliberals, and they were determined to "open" markets everywhere. They were out, as Thomas Friedman put it (though he hardly meant it quite this way), to flatten the Earth, which turned out to be a violent proposition.

The neoliberals were let loose to do their damnedest in the good times of the post–Cold War Clinton years. They wanted to apply a kind of American economic clout that they thought would never end to the organization of the planet. They believed the United States to be the economic superpower of the ages, and they had their own dreamy version of what an economic Pax Americana would be like. Privatization was the name of the game, and their version of shock-and-awe tactics involved calling in institutions like the International Monetary Fund to "discipline" developing countries into a profitable kind of poverty and misery.

In the end, gleefully slicing and dicing subprime mortgages, they "financialized" the world and so drove a hole through it. They were our economic jihadis, and in the great meltdown of 2008, they deep-sixed the world economy they had helped "unify." In the process, by increasing the gap between the superrich and everyone else, they helped create the 1 percent and the 99 percent in the United States and globally, preparing the ground for the protests to follow.

THE SECOND OCCUPATION

If the first occupation drove an economic stake through the heart of the planet, the second did a similar thing militarily. In the wake of the 9/11 attacks, the "unilateralists" of the Bush administration staked their own claim to a global occupation at the point of a cruise missile. Romantics all when it came to the U.S. military and what it could do, they invaded Iraq, determined to garrison the oil heartlands of the planet. It was going to be "shock and awe" and "mission accomplished" all the way. What they had in mind was a militarized version of an "occupy the wealth" scheme. Their urge to privatize even extended to the military itself, and when they invaded, in their baggage train came crony corporations ready to feast.

Once upon a time, Americans knew that only the monstrous enemy—most recently that "evil empire," the Soviet Union—could dream of world conquest and occupation. That was, by nature, what evil monsters did. Until 2001, when it turned out to be quite okay for the good guys of planet Earth to think along exactly the same lines.

The invasion of Iraq, that "cakewalk," was meant to establish a multigenerational foothold in the greater Middle East, including permanent bases garrisoned with thirty thousand to forty thousand American troops, and that was to be just the beginning of a chain reaction. Soon enough Syria and Iran would bow down before U.S. power or, if they refused, would go down anyway thanks to American techno-might. In the end, the lands of the greater Middle East would fall into line (with the help of Washington's proxy in the region, Israel).

And since there was no other nation or bloc of nations with anything like such military power, nor would any be allowed to arise, the result—and they weren't shy about this—would be a global Pax Americana and a domestic Pax Republicana more or less till the end of time. As the "sole superpower" or even "hyperpower," Washington would, in other words, occupy the planet.

Of course, Iraq and Afghanistan were also more traditional occupations. In Baghdad, for instance, American Consul L. Paul Bremer III issued Order 17, which essentially granted to every foreigner connected with the occupation enterprise the full freedom of the land, not to be interfered with in any way by Iraqis or any Iraqi political or legal institutions. This included "freedom of movement without delay throughout Iraq," and neither their vessels, vehicles, nor aircraft were to be "subject to registration, licensing, or inspection by the [Iraqi] Government." Nor in traveling would any foreign diplomat, soldier, consultant, or security guard, or any of their vehicles, vessels, or planes be subject to "dues, tolls, or charges, including landing and parking fees." And that was only the beginning.

Order 17, which read like an edict plucked directly from a nineteenth-century colonial setting, caught the local hubris of those privatizing occupiers.

All of this proved to be fantasy bordering on delusion, and it didn't take long for that to become apparent. In fact, the utter failure of the unilateralists came home to roost in the form of a status of forces (SOFA) agreement with Iraqi authorities that promised to end the U.S. garrisoning of the country not in 2030 or 2050, but in 2011. And the Bush administration felt forced to agree to it in 2008, the same year that the economic unilateralists were facing the endgame of their dreams of global domination.

In that year, the neoliberal effort to privatize the planet went down in flames, along with Lehman Brothers, all those subprime mortgages and derivatives, and a whole host of banks and financial outfits rescued from the trash bin of history by the U.S. Treasury. Talk about giving the phrase "creative destruction" the darkest meaning possible: the two waves of American unilateralists nearly took down the planet.

They let loose demons of every sort, even as they ensured that the world's first experience of a sole superpower would prove short indeed. Heap onto the rubble they left behind the global disaster of rising prices for the basics—food and fuel—and you have a situation so combustible that no one should have been surprised when a single Tunisian match set it aflame.

The first two failed occupations plunged the planet into chaos and misery, even as they paved the way, in a thoroughly unintended fashion, for an Arab Spring ready to take on the Middle East's 1 percent.

Note as well that, as their policies went to hell in a handbasket, the first and second set of occupiers walked off with their treasure and their selves intact. Neither the bankers nor the militarists went to jail, not a one of them. They had made out like bandits and continue to do so. They took home their multi-million-dollar bonuses. They kept their yachts, mansions, and (untaxed) private jets. They took with them the ability to sign million-dollar contracts for bestselling memoirs and to go on the lecture circuit at $100,000 to $150,000 a pop. They had, in the case of the second occupation, quite literally gotten away with murder (and torture and kidnapping, etc.). In the process, the misery of the 99 percent had been immeasurably increased.

THE THIRD OCCUPATION

The most significant and surprising thing the first two globalizing occupations did, however, was to globalize protest. Together they created the basis, in pure iniquity and inequity, in dead bodies and bruised lives, for Tahrir Square and Occupy Wall Street. Their failures set the stage for something new in the world.

The result was a Chalmers Johnson–style case of blowback, the spirit of which was caught in the protesters' appropriation of the very word *occupy*. There was a sense out there that they had occupied us long and disastrously enough. It was time for us to occupy them, and so our own parks, squares, streets, towns, cities, and countries.

The urge to right things is, in fact, a powerful one. Gene Turitz, a friend of mine who took part in the demonstrations that briefly shut down the port of Oakland, California, recently wrote me the following about the experience. It catches something of the mood of this moment:

> The mayor of Oakland, a former progressive, blasted the economic violence that was being perpetrated by the Occupy movement shutting down the port. No word about the economic violence of banks stealing people's homes through foreclosures, or the economic violence of [sports] team owners demanding the city build new stadiums for their teams or they will move to another city, or of corporations threatening to move if this or that is not done for them. That's just the way things are done. You do not want the "violence" of thousands of people peacefully showing that things must change to make their lives better.

Or in two words: we exist! And possibly in the nick of time.

THE FOURTH OCCUPATION

This is both the newest and oldest of occupations. I'm speaking about humanity's occupation of Earth. In recent centuries, can there be any question that we've been hard on this planet, exploiting it for everything it's worth? Our excuse was that we genuinely didn't know better, at least when it came to climate change, that we just didn't understand what kind of long-term harm the burning of fossil fuels could do. Now, of course, we know. Those who don't are either in denial or simply couldn't care less.

And here's just a taste of what we do know about how the fourth occupation is affecting the planet: Thirteen of the warmest years since record keeping began have occurred in the last fifteen years. In 2010, historically staggering amounts of carbon dioxide were sent into the atmosphere ("the biggest jump ever seen in global warming gases"). Extreme weather was, well, remarkably extreme in 2011—torrid droughts, massive fires, vast floods—and in the Arctic, ice is now melting at unprecedented rates, which will mean future sea-level rises that will threaten low-lying areas of the planet. And as for that temperature, well, it's going to keep going up, uncomfortably so.

Potentially, this is the monster blowback story of all time.

And here's just a taste of what we know about business as usual on this planet: If we rely on the previous occupiers and their ilk to save us, then it's going to be a long, dismal wait. Don't count on energy giants like Exxon or BP or their lobbyists and the politicians they influence to stop climate change. After all, none of them are going to be alive to see a far less habitable planet, so what do they care? Torrid zones are so then, profit sheets and bonuses are so now, which means: don't count on the 1 percent to give a damn.

If it were up to them—a few outliers among them excepted—we could probably simply write the Earth off as a future friendly place for us. And the planet wouldn't care. Give it one hundred thousand, 10 million, 100 million years, and it'll get itself back in shape with plenty of life-forms to go around.

We're such ephemeral creatures with such brief life spans. It's hard for us to think even in the sort of modestly long-range way that climate change demands. So thank your lucky stars that the first- and second-wave occupiers created a third payback occupation they never imagined possible. And thank your lucky stars that movements to occupy our planet in a new way and turn back the global warmers are slowly rising as well.

Like the attempted occupations of the global economy and the greater Middle East, each spurred by a sense of greed that went beyond all bounds, the occupation of our planet is guaranteed to create its own oppositional forces, and not just in the natural world either. They are perhaps already emerging along with the Arab Spring, the European summer, and the American fall, not to speak of the Russian winter. And when they're here—as the fifth occupation of planet Earth—when they stand their ground and chant, "We exist!" in anger, strength, and wonder, maybe then we can really tackle climate change and hope it isn't too late.

Maybe the fifth occupation is the one we're waiting for—and don't for a second doubt that it will come. It's already on its way.

APPENDIX B

EVIDENCE OF COSMIC COMMUNITY

Finley Eversole, Ph.D.

PART I. DISINFORMATION ABOUT ETS

I know for sure we're not alone in the universe. I have been privileged to be in on the fact that we've been visited on this planet. It's been covered up by our governments for sixty years now.

EDGAR MITCHELL, APOLLO ASTRONAUT
SIXTH MAN TO WALK ON THE MOON

It's the discovery of the life of humankind isn't it—to find out that we are not here alone.

PROFESSOR ROBERT JACOBS,
USAF—VANDENBERG AIR FORCE BASE

At no time, when the astronauts were in space were they alone: there was a constant surveillance by UFOs.

SCOTT CARPENTER, U.S. ASTRONAUT,
THE FOURTH MAN IN SPACE

315

> *For many years I have lived with a secret, in a secrecy imposed on all specialists in astronautics. I can now reveal that every day, in the USA, our radar instruments capture objects of form and composition unknown to us. And there are thousands of witness reports and a quantity of documents to prove this, but nobody wants to make them public.*
>
> MAJOR GORDON COOPER,
> U.S. ASTRONAUT

> *It is a truth the entire world has to be informed about, and that truth is that man is not alone.*
>
> SERGEANT CLIFFORD STONE,
> U.S. EXTRATERRESTRIAL RETRIEVAL TEAM

Campaigns of disinformation dating back to the 1940s have been employed by governments and the media to cover up one of the most significant events of our time—the recognition that humanity is not alone in the universe. The implications are profound and far-reaching. It may be worth noting that, contrary to media and mythic portrayals, there has never been a single incident of ET hostility toward Earth or its peoples. But they do seem concerned that we may destroy ourselves.

We already know that "black" military-governmental-industrial-intelligence programs have advanced energy and propulsion technologies reverse-engineered from downed ET craft and that these technologies would end forever our reliance on environmentally destructive fossil fuels and nuclear energy. The microchip, which gave us the PC, Internet, cell phone, and many other advances, is but one of several examples of reverse-engineered ET technology already in use. Given the severity of our energy and environmental problems, one must ask why this technology remains secret.

In an era when humans have traveled in space, why should we doubt that other self-conscious intelligent beings, perhaps far more advanced technologically and spiritually than ourselves, have also done so? And what are the amazing crop circles appearing all over the world, distinguished by brilliant geometric and astronomical messages, if not "calling cards" from our cosmic cousins—messages saying, "We are here to help if you will accept it."

As Steven Greer says:

> The recognition that mankind is one, that race, nationality, gender, religion and so on are secondary to our shared humanness, may well be the crowning achievement of the 20th century. . . . Our deepest point of unity transcends race, culture, gender, profession, life roles, even level of intelligence or emotional make-up, since all these attributes vary widely among people. Rather, the foundation of human oneness is consciousness itself. . . . All other human qualities arise from this mother of all attributes. . . . The finest, most enduring and transcendent foundation on which human unity is based then is consciousness itself; for we are all sentient beings, conscious, self-aware and intelligent. No matter how diverse two people or two cultures may be, this foundation of consciousness will enable unity to prevail, as it is the simplest yet most profound common ground which all humans share.
>
> The term Extraterrestrial Intelligence (ETI), so curiously nondescript, wonderfully lends itself to these concepts of unity. Regardless of planet, star system or galaxy of origin, and no matter how diverse, ETs are essentially intelligent, conscious, sentient beings. We are, essentially, one. On this basis, we may speak of one people inhabiting one universe, just as we now envision one people as children of one planet. Differences are always a matter of degree, but true unity established in consciousness is absolute. . . . For there is one universe inhabited by one people, and we are they.

REFERENCES

Books and Periodicals

Bamford, James. *The Puzzle Palace: A Report on America's Most Secret Agency.* Penguin, 1983.

———. *The Shadow Factory: The Ultra-Secret NSA from 9/11 to the Eavesdropping on America.* Doubleday, 2008.

Bearden, Thomas E. *Oblivion: America at the Brink.* Cheniere Press, 2005.

Corso, Philip. *The Day after Roswell.* Pocket Books, 1997.

Dolan, Richard M. *UFOs and the National Security State: Chronology of a Coverup, 1941–1973*. Hampton Roads Publishing, 2002.

Good, Timothy. *Above Top Secret*. Quill, 1989.

Goswami, Amit. *The Self-Aware Universe*. Tarcher/Putnam, 1993.

Greer, Steven M. *Extraterrestrial Contact: The Evidence and Implications*. Crossing Point Inc., 1999.

———. *Disclosure: Military and Government Witnesses Reveal the Greatest Secrets in Modern History*. Crossing Point Inc., 2001.

———. *Hidden Truth: Forbidden Knowledge*. Crossing Point Inc., 2006.

———. *Contact: Countdown to Transformation*. Crossing Point Inc., 2009.

Haines, Richard F. *CE-5: Close Encounters of the Fifth Kind*. Sourcebooks Inc., 1999.

Hellyer, Paul. *The Light at the End of the Tunnel*. AuthorHouse, 2010, 39–79. A retired U.S. Air Force general confirmed to former Canadian Defense Minister, Hon. Paul Hellyer, that everything in Philip Corso's book *The Day after Roswell* is true, and more.

Hill, Paul R. *Unconventional Flying Objects*. Hampton Roads Publishing, 1995.

Jahn, Robert G., and Brenda J. Dunne. *Margins of Reality: The Role of Consciousness in the Physical World*. Mariner Books, 1989.

Mack, John E. *Passport to the Cosmos*. White Crow Books, 2011.

Pursglove, David. *Zen in the Art of Close Encounters*. New Being Project, 1995.

Randel, Kevin D., and Donald R. Schmitt, *The Truth About the UFO Crash at Roswell*. Avon Books, 1994.

Ruppelt, Edward J. *The Report on Unidentified Flying Objects*. CreateSpace Independent Publishing Platform, 2012.

Salla, Michael E. *Exopolitics: Political Implications of the Extraterrestrial Presence*. Dandelion Books, 2004.

Wood, Ryan. *Majic Eyes Only*. Wood Enerprises, 2005.

Websites

Princeton Engineering Anomalies Research, www.princeton.edu/~pear/index.html (accessed June 18, 2012).

DVDs

The Disclosure Project. Two hours of witness testimony and briefing document; a distillation of over 120 hours of video interviews made by Dr Greer. Dozens of highly credible military and government witnesses discuss UFO events and projects they have worked on, with introduction and overview commentary by Dr. Greer. This work is a basic primer for those who wish to bridge the credibility gap once and for all. Available at www.DisclosureProject.org (888-382-7384).

The Disclosure Project. Four hours of witness testimony containing over three and a half hours, created as a Special Congressional Briefing provided to members of Congress, the executive branch, NASA officials, and senior military/Pentagon personnel. Available at www.DisclosureProject.org (888-382-7384)

Videos

Close Encounters: Proof of Alien Contact, www.snagfilms.com/films/title/close_encounters_proof_of_alien_contact (accessed June 18, 2012).

National Press Club, May 2001. Proceedings of historic National Press Club presentation of witnesses of UFOs. www.youtube.com/watch?v=nPgFBdvqC04 (accessed August 16, 2012).

PART II. CROP CIRCLES—COSMIC ART

God geometrizes.

PLATO

Why is the art world unmoved? This must surely be the most astonishing body of earth art in history. Why has academia not taken note of the numerological and geometrical skills demonstrated here? Why has the media not remarked on the perseverance and proficiency of this team of gifted operators?

MICHAEL GLICKMAN,

ARCHITECT

For me, it was immediately and ineluctably clear that these beautiful and joyous creations not only transcend our ordinary understanding of the material world, but confront our sense of the spiritual, too.

MIKE LEIGH,

FILM DIRECTOR

Any member of the press or public who still views crop circles as man-made hoaxes either has not studied the phenomenon or must wonder why some of the most brilliant and creative minds on the planet are not teaching in our colleges and universities. Apart from the incredible beauty of the designs, their amazing geometry, profound global symbolism, scientific and mathematical revelations, and the stunning speed with which they often appear (within fifteen minutes or less), there is the fact that plant stems are never broken and seeds from crops within crop circles germinate significantly faster and yield 61 percent more harvest. This is no ordinary phenomenon. No human has yet been able to produce a "genuine" crop circle in the presence of onlookers or the media. Crop circles, or "agriglyphs," are the work of a superior, moral, and compassionate intelligence that seems to want to communicate with humanity.

The continued appearance of crop circles year after year in more than seventy countries—now numbering several thousand—and their increasing richness of design should tell us we are not dealing with a handful of pranksters. Something significant is taking place. Perhaps these designs—impressed in the plant life of Mother Nature—are also meant to call us back to the Earth at a time when environmental problems threaten the life of humanity and allied kingdoms. Contempt prior to examination is the death of knowledge. There is a wisdom here we ought not ignore.

Julia Set crop circle, at Milk Hill, UK, is over eight hundred feet in diameter, contains 409 individual circles, and made its appearance during one of the rainiest nights of the year. Photo © Lucy Pringle, August 11, 2001.

Roundway Hill crop circle, Wiltshire, UK. It is an example of the beautiful geometry of many crop circles. Photo © Olivier Morel, July 25, 2010.

Human Butterfly—Metamorphosis crop circle near Goes, Netherlands. It is the largest crop circle ever! Its message: "Humans are evolving." August 7, 2009.

Cley Hill, Warminster, UK. It is a 3-D, six-sided cross within a hexagon. July 9, 2010.

Quetzalcoatl Native Headdress, Mayan 2012 crop circle, Silbury Hill, UK.
Photo © Steve and Karen Alexander, July 5, 2009.

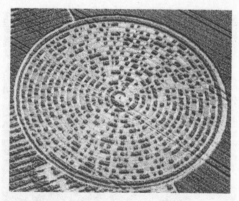

Detail of Sparsholt glyph, Hampshire, UK, 2002. The binary code message
reads, "BEWARE OF THE BEARERS OF FALSE GIFTS AND THEIR
BROKEN PROMISES. MUCH PAIN. BUT STILL TIME. BELIEVE.
THERE IS GOOD OUT THERE. WE OPPOSE DECEPTION.
CONDUIT CLOSING."

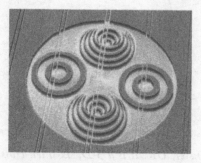

Silbury Hill crop circle, UK, 2001. Concentric and pyramiding circles.

*Crooked Soley crop circle, Berkshire, UK. Photo © Steve and
Karen Alexander, August 28, 2002.*

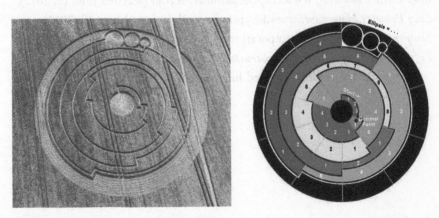

*Barbury Castle crop circle, UK, represents the value of π, symbolizing the
marriage of the finite and the infinite. Photo © Steve and Karen Alexander,
June 1, 2008.*

REFERENCES

Books and Periodicals

Alexander, Steve, and Karen Alexander. *Crop Circles: Signs, Wonders, and Mysteries*. Chartwell Books, 2006.

Glickman, Michael. *Crop Circles: The Bones of God*. Frog Books, 2009.

Howe, Linda Moulton. *Mysterious Lights and Crop Circles*. Pioneer Printing, 2002.

Silva, Freddy. *Secrets in the Fields: The Science and Mysticism of Crop Circles*. Hampton Roads Publishing, 2002.

Thomas, Andy. *Vital Signs*. Frog, Ltd., 2002.

DVDs

Crop Circles: Quest for Truth. Shout! Factory Theatre, 2002. Director: William Gazecki, featuring Colin Andrews.

Star Dreams. Genesis Communications, 2007. Director: Robert L. Nichol.

What on Earth? Inside the Crop Circle Mystery. Mighty Companions Inc., 2009. Director: Suzanne Taylor. For a preview of this DVD, go to: www.crop circlemovie.com (accessed June 18, 2012).

Websites

Crop Circle Connector, www.cropcircleconnector.com (accessed June 18, 2012).

Lucy Pringle's Homepage, www.lucypringle.co.uk (accessed June 18, 2012).

Temporary Temples, www.temporarytemples.co.uk (accessed June 18, 2012).

Zef Damen Crop Circle Reconstructions, www.zefdamen.nl/CropCircles/en/Crop_circles_en.htm (accessed June 18, 2012).

APPENDIX C

THE EARTH CHARTER

THE ORIGIN AND PURPOSE
OF THE EARTH CHARTER

The idea of the Earth Charter originated in 1987 when the United Nations World Commission on Environment and Development called for a charter that would guide humanity toward sustainable development. In 1992, then-UN Secretary General Boutros Boutros-Ghali brought the matter to the Rio de Janeiro Earth Summit, but it was felt that the time was not yet ripe for such a declaration. In 1994, Maurice Strong, chairman of the Earth Summit, and Mikhail Gorbachev, working through the Earth Council and Green Cross International respectively, restarted the Earth Charter as a civil society initiative with the help of the government of the Netherlands.

The Earth Charter is an international declaration of fundamental values and principles considered useful by its supporters for building a just, sustainable, and peaceful global society in the twenty-first century. Created by a global consultation process, and endorsed by organizations representing millions of people, the Charter "seeks to inspire in all peoples a sense of global interdependence and shared responsibility for the well-being of the human family, the greater community of life, and future generations." It calls upon humanity to cooperate in creating a global partnership based on the Earth Charter's ethical vision encompassing environmental protection, human rights, equitable human development, and peace among all peoples.

The *Earth Charter Initiative* exists to promote the Charter.

PREAMBLE

We stand at a critical moment in Earth's history, a time when humanity must choose its future. As the world becomes increasingly interdependent and fragile, the future at once holds great peril and great promise. To move forward we must recognize that in the midst of a magnificent diversity of cultures and life forms we are one human family and one Earth community with a common destiny. We must join together to bring forth a sustainable global society founded on respect for nature, universal human rights, economic justice, and a culture of peace. Towards this end, it is imperative that we, the peoples of Earth, declare our responsibility to one another, to the greater community of life, and to future generations.

EARTH, OUR HOME

Humanity is part of a vast evolving universe. Earth, our home, is alive with a unique community of life. The forces of nature make existence a demanding and uncertain adventure, but Earth has provided the conditions essential to life's evolution. The resilience of the community of life and the well-being of humanity depend upon preserving a healthy biosphere with all its ecological systems, a rich variety of plants and animals, fertile soils, pure waters, and clean air. The global environment with its finite resources is a common concern of all peoples. The protection of Earth's vitality, diversity, and beauty is a sacred trust.

THE GLOBAL SITUATION

The dominant patterns of production and consumption are causing environmental devastation, the depletion of resources, and a massive extinction of species. Communities are being undermined. The benefits of development are not shared equitably and the gap between rich and poor is widening. Injustice, poverty, ignorance, and violent conflict are widespread and the cause of great suffering. An unprecedented rise in human population has overburdened ecological and social systems. The foundations of global security are threatened. These trends are perilous—but not inevitable.

THE CHALLENGES AHEAD

The choice is ours: form a global partnership to care for Earth and one another or risk the destruction of ourselves and the diversity of life. Fundamental changes are needed in our values, institutions, and ways of living. We must realize that when basic needs have been met, human development is primarily about being more, not having more. We have the knowledge and technology to provide for all and to reduce our impacts on the environment. The emergence of a global civil society is creating new opportunities to build a democratic and humane world. Our environmental, economic, political, social, and spiritual challenges are interconnected, and together we can forge inclusive solutions.

UNIVERSAL RESPONSIBILITY

To realize these aspirations, we must decide to live with a sense of universal responsibility, identifying ourselves with the whole Earth community as well as our local communities. We are at once citizens of different nations and of one world in which the local and global are linked. Everyone shares responsibility for the present and future well-being of the human family and the larger living world. The spirit of human solidarity and kinship with all life is strengthened when we live with reverence for the mystery of being, gratitude for the gift of life, and humility regarding the human place in nature.

We urgently need a shared vision of basic values to provide an ethical foundation for the emerging world community. Therefore, together in hope we affirm the following interdependent principles for a sustainable way of life as a common standard by which the conduct of all individuals, organizations, businesses, governments, and transnational institutions is to be guided and assessed.

PRINCIPLES

I. RESPECT AND CARE FOR THE COMMUNITY OF LIFE

1. Respect Earth and life in all its diversity.

   a. Recognize that all beings are interdependent and every form of life has value regardless of its worth to human beings.

   b. Affirm faith in the inherent dignity of all human beings and in the intellectual, artistic, ethical, and spiritual potential of humanity.

2. Care for the community of life with understanding, compassion, and love.

   a. Accept that with the right to own, manage, and use natural resources comes the duty to prevent environmental harm and to protect the rights of people.

   b. Affirm that with increased freedom, knowledge, and power comes increased responsibility to promote the common good.

3. Build democratic societies that are just, participatory, sustainable, and peaceful.

   a. Ensure that communities at all levels guarantee human rights and fundamental freedoms and provide everyone an opportunity to realize his or her full potential.

   b. Promote social and economic justice, enabling all to achieve a secure and meaningful livelihood that is ecologically responsible.

4. Secure Earth's bounty and beauty for present and future generations.

   a. Recognize that the freedom of action of each generation is qualified by the needs of future generations.

b. Transmit to future generations values, traditions, and institutions that support the long-term flourishing of Earth's human and ecological communities.

In order to fulfill these four broad commitments, it is necessary to:

II. ECOLOGICAL INTEGRITY

5. **Protect and restore the integrity of Earth's ecological systems, with special concern for biological diversity and the natural processes that sustain life.**
   a. Adopt at all levels sustainable development plans and regulations that make environmental conservation and rehabilitation integral to all development initiatives.
   b. Establish and safeguard viable nature and biosphere reserves, including wild lands and marine areas, to protect Earth's life support systems, maintain biodiversity, and preserve our natural heritage.
   c. Promote the recovery of endangered species and ecosystems.
   d. Control and eradicate non-native or genetically modified organisms harmful to native species and the environment, and prevent introduction of such harmful organisms.
   e. Manage the use of renewable resources such as water, soil, forest products, and marine life in ways that do not exceed rates of regeneration and that protect the health of ecosystems.
   f. Manage the extraction and use of non-renewable resources such as minerals and fossil fuels in ways that minimize depletion and cause no serious environmental damage.

6. **Prevent harm as the best method of environmental protection and, when knowledge is limited, apply a precautionary approach.**
   a. Take action to avoid the possibility of serious or irreversible environmental harm even when scientific knowledge is incomplete or inconclusive.

b. Place the burden of proof on those who argue that a proposed activity will not cause significant harm, and make the responsible parties liable for environmental harm.

c. Ensure that decision making addresses the cumulative, long-term, indirect, long distance, and global consequences of human activities.

d. Prevent pollution of any part of the environment and allow no build-up of radioactive, toxic, or other hazardous substances.

e. Avoid military activities damaging to the environment.

7. **Adopt patterns of production, consumption, and reproduction that safeguard Earth's regenerative capacities, human rights, and community well-being.**

a. Reduce, reuse, and recycle the materials used in production and consumption systems, and ensure that residual waste can be assimilated by ecological systems.

b. Act with restraint and efficiency when using energy, and rely increasingly on renewable energy sources such as solar and wind.

c. Promote the development, adoption, and equitable transfer of environmentally sound technologies.

d. Internalize the full environmental and social costs of goods and services in the selling price, and enable consumers to identify products that meet the highest social and environmental standards.

e. Ensure universal access to health care that fosters reproductive health and responsible reproduction.

f. Adopt lifestyles that emphasize the quality of life and material sufficiency in a finite world.

8. **Advance the study of ecological sustainability and promote the open exchange and wide application of the knowledge acquired.**

a. Support international scientific and technical cooperation on

sustainability, with special attention to the needs of developing nations.

b. Recognize and preserve the traditional knowledge and spiritual wisdom in all cultures that contribute to environmental
protection and human well-being.

c. Ensure that information of vital importance to human health
and environmental protection, including genetic information, remains available in the public domain.

III. SOCIAL AND ECONOMIC JUSTICE

9. **Eradicate poverty as an ethical, social, and environmental
    imperative.**

a. Guarantee the right to potable water, clean air, food security,
uncontaminated soil, shelter, and safe sanitation, allocating
the national and international resources required.

b. Empower every human being with the education and
resources to secure a sustainable livelihood, and provide
social security and safety nets for those who are unable to
support themselves.

c. Recognize the ignored, protect the vulnerable, serve those
who suffer, and enable them to develop their capacities and
to pursue their aspirations.

10. **Ensure that economic activities and institutions at all levels
    promote human development in an equitable and sustain-
    able manner.**

a. Promote the equitable distribution of wealth within nations
and among nations.

b. Enhance the intellectual, financial, technical, and social
resources of developing nations, and relieve them of onerous
international debt.

c. Ensure that all trade supports sustainable resource use, environmental protection, and progressive labor standards.

d. Require multinational corporations and international financial organizations to act transparently in the public good,

and hold them accountable for the consequences of their activities.

11. **Affirm gender equality and equity as prerequisites to sustainable development and ensure universal access to education, health care, and economic opportunity.**

   a. Secure the human rights of women and girls and end all violence against them.

   b. Promote the active participation of women in all aspects of economic, political, civil, social, and cultural life as full and equal partners, decision makers, leaders, and beneficiaries.

   c. Strengthen families and ensure the safety and loving nurture of all family members.

12. **Uphold the right of all, without discrimination, to a natural and social environment supportive of human dignity, bodily health, and spiritual well-being, with special attention to the rights of indigenous peoples and minorities.**

   a. Eliminate discrimination in all its forms, such as that based on race, color, sex, sexual orientation, religion, language, and national, ethnic or social origin.

   b. Affirm the right of indigenous peoples to their spirituality, knowledge, lands and resources and to their related practice of sustainable livelihoods.

   c. Honor and support the young people of our communities, enabling them to fulfill their essential role in creating sustainable societies.

   d. Protect and restore outstanding places of cultural and spiritual significance.

IV. DEMOCRACY, NONVIOLENCE, AND PEACE

13. **Strengthen democratic institutions at all levels, and provide transparency and accountability in governance, inclusive participation in decision making, and access to justice.**

   a. Uphold the right of everyone to receive clear and timely information on environmental matters and all development plans and activities which are likely to affect them or in which they have an interest.

b. Support local, regional and global civil society, and promote the meaningful participation of all interested individuals and organizations in decision making.

c. Protect the rights to freedom of opinion, expression, peaceful assembly, association, and dissent.

d. Institute effective and efficient access to administrative and independent judicial procedures, including remedies and redress for environmental harm and the threat of such harm.

e. Eliminate corruption in all public and private institutions.

f. Strengthen local communities, enabling them to care for their environments, and assign environmental responsibilities to the levels of government where they can be carried out most effectively.

14. **Integrate into formal education and life-long learning the knowledge, values, and skills needed for a sustainable way of life.**

a. Provide all, especially children and youth, with educational opportunities that empower them to contribute actively to sustainable development.

b. Promote the contribution of the arts and humanities as well as the sciences in sustainability education.

c. Enhance the role of the mass media in raising awareness of ecological and social challenges.

d. Recognize the importance of moral and spiritual education for sustainable living.

15. **Treat all living beings with respect and consideration.**

a. Prevent cruelty to animals kept in human societies and protect them from suffering.

b. Protect wild animals from methods of hunting, trapping, and fishing that cause extreme, prolonged, or avoidable suffering.

    c. Avoid or eliminate to the full extent possible the taking or destruction of non-targeted species.

16. Promote a culture of tolerance, nonviolence, and peace.

    a. Encourage and support mutual understanding, solidarity, and cooperation among all peoples and within and among nations.

    b. Implement comprehensive strategies to prevent violent conflict and use collaborative problem solving to manage and resolve environmental conflicts and other disputes.

    c. Demilitarize national security systems to the level of a nonprovocative defense posture, and convert military resources to peaceful purposes, including ecological restoration.

    d. Eliminate nuclear, biological, and toxic weapons and other weapons of mass destruction.

    e. Ensure that the use of orbital and outer space supports environmental protection and peace.

    f. Recognize that peace is the wholeness created by right relationships with oneself, other persons, other cultures, other life, Earth, and the larger whole of which all are a part.

THE WAY FORWARD

As never before in history, common destiny beckons us to seek a new beginning. Such renewal is the promise of these Earth Charter principles. To fulfill this promise, we must commit ourselves to adopt and promote the values and objectives of the Charter.

    This requires a change of mind and heart. It requires a new sense of global interdependence and universal responsibility. We must imaginatively develop and apply the vision of a sustainable way of life locally, nationally, regionally, and globally. Our cultural diversity is a precious heritage and different cultures will find their own distinctive ways to realize the vision. We must deepen and expand the global dialogue that generated the Earth Charter, for we have much to learn from the ongoing collaborative search for truth and wisdom.

Life often involves tensions between important values. This can mean difficult choices. However, we must find ways to harmonize diversity with unity, the exercise of freedom with the common good, short-term objectives with long-term goals. Every individual, family, organization, and community has a vital role to play. The arts, sciences, religions, educational institutions, media, businesses, nongovernmental organizations, and governments are all called to offer creative leadership. The partnership of government, civil society, and business is essential for effective governance.

In order to build a sustainable global community, the nations of the world must renew their commitment to the United Nations, fulfill their obligations under existing international agreements, and support the implementation of Earth Charter principles with an international legally binding instrument on environment and development.

Let ours be a time remembered for the awakening of a new reverence for life, the firm resolve to achieve sustainability, the quickening of the struggle for justice and peace, and the joyful celebration of life.

NOTES

CHAPTER 1. NIKOLA TESLA:
ELECTRICAL SAVANT

1. Nikola Tesla, *My Inventions: The Autobiography of Nikola Tesla*, ed. Ben Johnston (Williston, Vt.: Hart Brothers Publishers, 1982), 41.

2. John O'Neill, *Prodigal Genius* (New York: Ives-Washburn, 1944), 107.

3. Nikola Tesla, "Experiments with Alternating Currents of Very High Frequency and Their Application to Methods of Artificial Illumination." Lecture, Columbia College [University], New York, May 20, 1891, in *Nikola Tesla (1856–1943) Lectures, Patents, and Articles* (Belgrade, Yugoslavia: Tesla Museum, 1956), L15–L47; Nikola Tesla, "Experiments with Alternate Currents of High Potential and High Frequency." Lecture delivered before the I.E.E., London, February 1892, in *Nikola Tesla (1856–1943) Lectures, Patents, and Articles* (Belgrade, Yugoslavia: Tesla Museum, 1956), L48–L106.

4. Nikola Tesla, *The Inventions, Researches and Writings of Nikola Tesla*, ed. T. C. Martin (New York: Electrical Engineer, 1894).

5. Lawrence Lessing, *Man of High Fidelity: Edwin Howard Armstrong* (Philadelphia, Pa.: Lippincott, 1956), 42–43; Marc Seifer, *Wizard: The Life and Times of Nikola Tesla* (New York: Citadel Press, 1997), 373–74.

6. Orrin Dunlap, *Radio's 100 Men of Science* (New York: Harper and Bros., 1944), 156–158.

7. Marc Seifer, *Wizard: The Life and Times of Nikola Tesla* (New York: Citadel Press, 1996), 464.

8. Nikola Tesla, "Experiments with Alternating Currents of Very High Frequency and Their Application to Methods of Artificial Illumination."

Lecture, Columbia College [University], New York, May 20, 1891, in *Nikola Tesla (1856–1943) Lectures, Patents, and Articles* (Belgrade, Yugoslavia: Tesla Museum, 1956), L15–L47.

9. Nikola Tesla, "Developments in Practice and Art of Telephotography," *Electrical Review*, December 11, 1920, in *Lectures, Patents, and Articles* (Belgrade, Yugoslavia: Tesla Museum, 1956), A94–A97.

10. Marc Seifer, *Wizard: The Life and Times of Nikola Tesla* (New York: Citadel Press, 1996), 236–303.

11. Franklin D. Roosevelt letter to the U.S. Navy, September 14, 1916. National Archives, Washington, DC.

12. Franklin D. Roosevelt letter to the U.S. Navy, September 14, 1916. National Archives, Washington, DC; Marconi Wireless vs. United States, U.S. Supreme Court, 320 U.S. 1, decided June 21, 1943.

13. November 15, 2003 e-mail to Marc Seifer from Klaus Jebens, son of Heinreich Jebens.

14. Nikola Tesla, "Dr. Tesla Writes of Various Phases of His Discovery," *New York Times*, February 6, 1932.

15. Nikola Tesla, "Experiments with Alternating Currents of Very High Frequency and Their Application to Methods of Artificial Illumination." Lecture, Columbia College [University], New York, May 20, 1891, in *Nikola Tesla (1856–1943) Lectures, Patents, and Articles* (Belgrade, Yugoslavia: Tesla Museum, 1956); Marc Seifer, *Wizard: The Life and Times of Nikola Tesla* (New York: Citadel Press, 1996), 473.

CHAPTER 2. JOHN WORRELL KEELY

1. Clara Bloomfield-Moore, *Keely and His Discoveries, Aerial Navigation*, (London: Kegan Paul, Trench Trübner & Co., 1893), 333.

2. "Keely's Secret," *The World*, May 11, 1890.This interview in *The World* is briefly mentioned in "The Keely Motor Secret," *The Atlanta Constitution*, May 21, 1890.

3. "Keely, Motor Man, Dead," *Public Ledger and Daily Transcript*, November 19, 1898. About Keely's mechanical turn of mind, see also short reference in: William Mill Butler, "Keely and the Keely Motor," *The Home Magazine*, 1898.

4. "Keely's Secret," *The World*, May 11, 1890. How Keely exposed the mediums he never explained, but it is possible that here we have the nucleus of the tales that Keely was connected to a circus, a sleight-of-hand performer or showed amazing dexterity with card tricks, since mediums often were—and

still are—exposed by magicians or stage conjurors. See: Carl Sifakis, *Hoaxes and Scams* (New York: Facts on File, 1994).

5. A Wilford Hall, "John Keely—A personal interview," *Scientific Arena*, January 1887. See also: William Mill Butler, "Keely and the Keely Motor," *The Home Magazine*, 1898.

6. "Count Von Rosen Talks of Keely," *The Evening Bulletin*, March 15, 1899.

7. "Noted Woman Dead," *Public Ledger and Daily Transcript*, January 6, 1899.

8. "The Keely Motor," *International Cyclopedia*, vol. VIII (1899): 458.

9. *The Keely Motor Criticized*, a republication in pamphlet form of a series of editorials that appeared in the *Public Record* of Philadelphia, August 3–6, 1875 (n.p., n.d., but in all probability published *Public Record* [1875]: 2–5).

10. Ibid., *Keely and His Discoveries, Aerial Navigation*, 10, 11, 320, 150.

11. Ibid., *The Keely Motor Criticized*, 2–5.

12. Ibid.

13. Ibid.

14. Undated (but probably around 1892) letter by C. G. Till of Brooklyn, New York. In: Clara Bloomfield-Moore, *Keely and His Discoveries, Aerial Navigation* (London: Kegan Paul, Trench, Trübner & Co., 1893), 320.

15. Megargee, "Seen and Heard in Many Places," *The Times*, March 11, 1898. A large part of the text was exactly repeated several months later in his "Seen and Heard in Many Places," *The Times*, November 21, 1898.

16. Ibid., *The Keely Motor Criticized*, 5.

17. William Mill Butler, "Keely and the Keely Motor," *The Home Magazine* (1898): 106.

18. Ibid., *The Keely Motor Criticized*, 5.

19. "The Keely Motor," *New York Times*, November 6, 1875.

20. "Keely, Motor Man Dead," *Public Ledger and Daily Transcript*, November 19, 1898.

21. "Patent Application of John Ernest Worrell Keely," *Sympathetic Vibratory Physics* (4) (12) (1989): 7–9.

22. Ibid., *The Keely Motor Criticized*, 2–5.

23. O. M. Babcock, *The Keely Motor, Financial, Mechanical, Philosophical, Historical, Actual, Prospective*, privately printed in Philadelphia (June 1881): 25. Facsimile reprinted by Delta Spectrum Research.

24. Ibid.

25. Clara Bloomfield-Moore, *Keely's Secrets*, T.P.S. (1888): 17. Also in Clara Bloomfield-Moore, *Keely and His Discoveries, Aerial Navigation* (London: Kegan Paul, Trench, Trübner & Co., 1893), 87.

26. Ibid.

27. Ibid., "Seen and Heard in Many Places."

28. "Fortune Was Squandered in Keely Motor," *The Press*, January 22, 1899.

29. Henry Stevens, *Hitler's Suppressed and Still-Secret Weapons, Science and Technology* (Kempton, Ill.: Adventures Unlimited Press, 2007), 209–211. In regard to Eric von Rosen, Hermann Göring, and the Edelweiss Gesellschaft, see Theo Paijmans, "House of the Swastika," *Darklore 7*, 2012.

30. "Did Another Share Keely's Secret? A Mysterious Stranger Whose Presence is a Puzzle to the Directors," *The Evening Bulletin*, November 26, 1898.

31. "Official Statement to the Journal by a Director," in: "Keely Monumental Fraud of the Century," *New York Journal*, January 29, 1899.

32. "Will Ship Keely's Devices," *Public Ledger and Daily Transcript*, December 28, 1898.

33. "Keely Inventory Filed," *The Evening Bulletin*, January 4, 1899.

34. "Keely's Motor in Boston," unspecified newspaper clipping, January 3, 1899, Sympathetic Vibratory Physics Homepage.

35. "Kinraide on Keely," *The Press*, July 16, 1899.

36. Ibid.

37. "Keely Motor Tricks Were All Reproduced," *The Press*, July 16, 1899.

38. Ibid.

39. Ibid.

40. Ibid.

CHAPTER 3. VIKTOR SCHAUBERGER

1. Viktor Schauberger, *Our Senseless Toil*. In *Nature as Teacher* (n.p.: Gateway Publishing, 1998), pt. I, 28–29 (see footnote 16).

2. Viktor Schauberger, "The First Biotechnical Practice" [Die erste biotechnische Praxis], *Implosion*, no. 7 (n.d.): 1; Viktor Schauberger, "Let the Upheaval Begin!" [Den Umbruch beginnen!], *Implosion*, no. 67 (1977): 1.

3. Viktor Schauberger, "Allgemeines zu Apparaten/Maschinen" or "General Data on Apparatuses/Machines," *Mensch und Technik* Year 24, vol. 2, section 7.4 (1993): 49. Special edition wholly devoted to recently discovered information on Viktor Schauberger contained in Arnold Hohls's notebook.

4. "DNA Double Helix" (image/diagram) in *The Molecular Biology of the Cell*, eds. Bruce Alberts, Dennis Bray, Julian Lewis, Martin Raff, Keith Roberts, and James D. Watson (New York: Garland Publishing Inc., 1983), 101.

5. From a list of Viktor Schauberger quotations in the Schauberger archives.

6. H. L. Penman, "The Water Cycle," in *The Biosphere: A Scientific American Book* (New York: W. H. Freeman and Co., 1970).

7. Much of this material is to be found translated into English in Callum Coats, ed. and trans., *The Water Wizard,* Ecotechnology Series, vol. 1 (Bath, U.K.: Gateway Books, 1998).

8. The full text to be found in Coats, *Water Wizard,* vol. 1.

9. Schauberger, *Our Senseless Toil,* pt. II, 6.

10. Data from Kenneth S. Davis and John Arthur Day, *Water—The Mirror of Science* (London: Heinemann Educational Books, 1964); and Claude A. Villee, Eldra P. Solomon, and P. William Davis, *Biology,* international ed. (Philadelphia: Saunders College Publishing, 1985).

11. *American Journal of Health* as reported in *The Australian* newspaper of July 2, 1992.

12. Viktor Schauberger, *Our Senseless Toil,* pt. II, 17.

CHAPTER 4.
ROYAL RAYMOND RIFE

1. R. E. Seidel and Elizabeth Winter. "The New Microscope." *Journal of the Franklin Institute* (237) (2) (1944).

2. For a fuller account of the role of Lawrence Fishbein and the American Medical Association in the suppression of Rife's extraordinary healing technology, see Barry Lynes, *The Cancer Cure That Worked!: Fifty Years of Suppression* (Ontario: Compcare Publications, 1986).

CHAPTER 5.
T. TOWNSEND BROWN

1. For additional information on T. Townsend Brown and antigravity research, see Paul A. LaViolette, Ph.D., *Secrets of Antigravity Propulsion: Tesla, UFOs, and Classified Aerospace Technology,* 3rd ed. (Rochester, Vt.: Bear & Company, 2008).

2. William Moore and Charles Berlitz, *The Philadelphia Experiment: Project Invisibility* (New York: Fawcett, 1995).

CHAPTER 6. THE SUSTAINABLE
TECHNOLOGY SOLUTION
REVOLUTION

1. Daniel Quinn, *Ishmael: An Adventure of the Mind and Spirit* (New York: Bantam, 1992).

2. Brian O'Leary, "The Turquoise Revolution," *Infinite Energy* (10) (93) (September/October 2010), also posted on www.brianoleary.info (accessed June 16, 2012) and www.drbrianoleary.wordpress.com/?s=turquoise+revolution (accessed June 16, 2012).

3. Gunter Pauli, *The Blue Economy: 10 Years, 100 Innovations, 100 Million Jobs* (Taos, N. Mex.: Paradigm Publications, 2010). More information is available at the Zero Emissions Research and Initiatives website, www.zeri.org (accessed June 16, 2012).

4. Patrick Kelly, "Practical Guide to Free Energy Devices," www.free-energy-info.com (accessed June 16, 2012). This website is an excellent review of the state of the art of new energy technologies, including descriptions of specific devices.

5. John Perkins, *Hoodwinked* (New York: Broadway Books, 2009).

6. Brian O'Leary, *Re-Inheriting the Earth: Awakening to Sustainable Solutions and Greater Truths* (published by Dr. Brian O'Leary, 2003).

7. www.exemplarzero.org (accessed July 3, 2012).

8. Brian O'Leary, *The Energy Solution Revolution*, 2nd ed. (n.p.: Bridger House, 2009). Also available on www.createspace.com/3407065 (accessed June 16, 2012) or as an e-book on www.brianoleary.info/ (accessed June 16, 2012).

9. Buckminster Fuller, as quoted on www.projectearth.com (accessed August 16, 2012).

10. Alick Bartholomew, *The Story of Water: Source of Life* (Edinburgh, UK: Floris Books, 2010). This book is a superb review of the special properties of water. Another excellent resource is the 2010 Russian documentary *Water: The Great Mystery*, now available on the Internet at: www.youtube.com/watch?v=kM8KtN-bqXA.

11. Gunter Pauli, *Blue Economy*, www.zeri.org (accessed June 16, 2012); James Aronson, Suzanne J. Milton, and James N. Blignaut, *Restoring Natural Capital: Science, Business, and Practice* (Washington, D.C.: Island Press, 2007); and the Society for Ecological Restoration International's website, www.ser.org (accessed June 16, 2012).

12. Eugene Linden, "Biodiversity: The Death of Birth," *Time*, June 24, 2001.

13. David Yurth, *The Ho Chi Minh Handbook of Guerilla Warfare: Strategies for Innovation Management* (forthcoming).

14. Gary Vesperman, "History of 'New Energy' Invention Suppression Cases," www.rense.com/general72/oinvent.htm (accessed June 16, 2012).

CHAPTER 8. IMAGINE A FREE ENERGY FUTURE FOR ALL OF HUMANITY

1. Steven M. Greer, *Disclosure: Military and Government Witnesses Reveal the Greatest Secrets in Modern History* (Crozet, Va.: Crossing Point Inc., 2001).

CHAPTER 9. ENERGY TECHNOLOGIES FOR THE TWENTY-FIRST CENTURY

1. Paul A. LaViolette, "Moving beyond the First Law and Advanced Field Propulsion Technologies," in *"Outside-the-Box" Technologies, Their Critical Role Concerning Environmental Trends, and the Unnecessary Energy Crisis,* ed. T. Loder, U.S. Senate Environment and Public Works Commission, http://epw.senate.gov/107th/loder.htm (accessed August 15, 2012).

2. T. T. Brown, "How I Control Gravity," *Science and Information Magazine,* August 1929. Reprinted in *Psychic Observer* (37) (1) (n.d.): 66–67.

3. Hermann Oberth, "Flying Saucers Come from a Distant World," *The American Weekly,* October 24, 1954.

4. Aviation Studies (International) Ltd., "Electro-gravitics Systems: An Examination of Electrostatic Motion, Dynamic Counterbary, and Barycentric Control," in *Electrogravitics Systems: Reports on a New Propulsion Methodology,* ed. Thomas Valone, 14 (Washington, D.C.: Integrity Research Institute, 1994).

5. Ibid., 27.

6. Ibid., 19.

7. Gravity Rand Ltd., "The Gravitics Situation," in *Electrogravitics Systems: Reports on a New Propulsion Methodology,* ed. Thomas Valone (Washington, D.C.: Integrity Research Institute, 1994), 54.

8. Aviation Studies (International) Ltd., "Electro-gravitics Systems" (1956): 11.

9. Ibid., 34.

10. Ibid., 41.

11. Gravity Rand Ltd., "Gravitics Situation," 47.

12. Aviation Studies (International) Ltd., "Electrogravitics Systems" (1956): 32.

13. M. W. Evans, "The Link between the Sachs and O(3) Theories of Electrodynamics," in *Modern Nonlinear Physics: Advances in Chemical Physics 19,* 2nd ed., ed. M. W. Evans (2002): 469–94.

14. P. K. Anastasovski, T. E. Bearden, C. Ciubotariu, et al., "Anti Gravity Effects in the Sachs Theory of Electrodynamics," *Foundations of Physics Letters* (14) (6) (2001): 601–5.

15. M. Alcubierre, "The Warp Drive: Hyper-fast Travel within General Relativity," *Classical and Quantum Gravity* 11 (1994): L73.

16. H. E. Puthoff, "SETI, the Velocity-of-Light Limitation, and the Alcubierre Warp Drive: An Integrating Overview," *Physics Essays* 9 (1996): 156.

17. S. K. Lamoreaux, "Demonstration of the Casmir Force in the 0.6 to 6 μm Range," *Physical Review Letters* 78 (1997): 5.

18. H. Puthoff, "Gravity as a Zero-Point Fluctuation Force," *Physical Review A* 39, no. 5 (1989): 2333–42; H. Puthoff, "Source of Electromagnetic Zero-Point Energy," *Physical Review A* 40, no. 9 (1989): 4597–4862.

19. See Thomas E. Bearden's website for an extensive listing and copies of his papers at www.cheniere.org (accessed June 16, 2012).

20. P. K. Anastasovski, T. E. Bearden, C. Ciubotariu, et al. "Explanation of the Motionless Electromagnetic Generator with 0(3) Electrodynamics," *Foundations of Physics Letters* (14) (1) (2001): 87–93.

21. Thomas E. Bearden, *Energy from the Vacuum: Concepts and Principles* (Santa Barbara, Calif.: Cheniere Press, 2002).

22. Thomas Valone, "The Right Time to Develop Future Energy Technologies," in "'Outside-the-Box' Technologies, Their Critical Role Concerning Environmental Trends, and the Unnecessary Energy Crisis," ed. T. Loder, http://theori onproject.org/en/righttime_valone.html (accessed August 15, 2012).

23. H. Puthoff, "Everything for Nothing," *New Scientist,* July 28, 1990, 52–55.

24. B. Haisch, A. Rueda, and H. Puthoff, "Beyond E = mc²: A First Glimpse of a Postmodern Physics, in which Mass, Inertia, and Gravity Arise from Underlying Electromagnetic Processes," *The Sciences* 34 (1994): 26; B. Haisch, A. Rueda, and H. Puthoff, "Physics of the Zero-Point Field: Implications for Inertia, Gravitation, and Mass," *Speculations in Science and Technology* 20 (1997): 99; B. Haisch and A. Rueda, "An Electromagnetic Basis for Inertia and Gravitation: What Are the Implications for 21st Century Physics and Technology?" in *Space Technology and Applications International Forum—1998,* ed. Mohamed S. El-Genk (1443); Bernhard Haisch and Alfonso Rueda, "The Zero-Point Field and the NASA Challenge to Create the Space Drive," in *Proceedings of the NASA Breakthrough Propulsion Physics Workshop* (1999): 55.

25. P. A. LaViolette, "The U.S. Antigravity Squadron," in *Electrogravitics Systems: Reports on a New Propulsion Methodology,* ed. Thomas Valone (Washington, D.C.: Integrity Research Institute, 1993), 82–101.

26. P. A. LaViolette, "Electrogravitics: Back to the Future," *Electric Spacecraft Journal,* no. 4 (1992): 23–28; P. A. LaViolette, "A Theory of Electrogravitics," *Electric Spacecraft Journal,* no. 8 (1993): 33–36.

27. LaViolette, "Moving Beyond the First Law."

28. Information available at the Disclosure Project website: www.disclosure-project.org (accessed June 16, 2012).

29. S. M. Greer and T. C. Loder III. "Disclosure Project Briefing Document." Available on CD from: The Disclosure Project, P.O. Box 2365, Charlottesville, Va., 22902. Also available from: www.disclosureproject.org (accessed June 16, 2012).

30. Steven M. Greer, *Disclosure: Military and Government Witnesses Reveal the Greatest Secrets in Modern History* (Crozet, Va.: Crossing Point Inc., 2001).

31. Ibid., 357–66.

32. Ibid., 262–70.

33. Ibid., 384–87.

34. Ibid., 391–403.

35. Ibid., 388–89.

36. Ibid., 497–510.

37. Nick Cook, *The Hunt for Zero Point: Inside the World of Antigravity* (New York: Broadway Books, 2002).

38. L. Deavenport, "T. T. Brown Experiment Replicated," *Electric Spacecraft Journal,* no. 16 (October 1995). Reprinted in *Electrogravitics Systems: Reports on a New Propulsion Methodology,* ed. Thomas Valone (Washington, D.C.: Integrity Research Institute, 1994).

39. Thomas B. Bahder and Chris Fazi, *Force on an Asymmetric Capacitor* (Adelphi, Md.: Army Research Laboratory, 2002). Available at: http://lifterproject.online.fr/arl_fac/index.html (accessed August 16, 2012).

40. Transdimensional Technologies, 906-E Bob Wallace Ave., Huntsville, Ala. 35801.

41. For more information, see the JLN Labs website at: http://jnaudin.free.fr (accessed June 16, 2012).

42. Greer and Loder, "Disclosure Project Briefing Document."

43. Ben Rich (lecture, University of California, Los Angeles Engineering Department, March 23, 1993).

44. Halton Arp, *Seeing Red: Redshifts, Cosmology, and Academic Science* (Montreal, Que.: Aperion, 1998), 249.

CHAPTER 11. COLD FUSION

1. B. Y. Liaw, P. L. Tao, and B. E. Liebert, "Helium Analysis of Palladium Electrodes after Molten Salt Electrolysis," *Fusion Science and Technology* 23 (1993): 92.

2. G. Mengoli, M. Bernardini, C. Manduchi, and G. Zannoni, "Calorimetry Close to the Boiling Temperature of the D_2O/Pd Electrolytic System," *Journal of Electroanalytical Chemistry* 444 (1998): 155.

3. Georges Lonchampt, Jean-Paul Siberian, Lucien Bonnetain, and Jean Delepine, "Excess Heat Measurement with Pons and Fleischmann Type Cells," in *The Seventh International Conference on Cold Fusion, Vancouver, Canada, 1998* (Salt Lake City, Utah: ENECO Inc., 1998).

4. M. Fleischmann and S. Pons, "Calorimetry of the Pd-D_2O System: From Simplicity via Complications to Simplicity," *Physics Letters A* 176 (1993): 118.

5. Edmund K. Storms, *The Science of Low Energy Nuclear Reaction: A Comprehensive Compilation of Evidence and Explanations about Cold Fusion* (Singapore: World Scientific, 2007), 312.

6. M. Fleischmann, S. Pons, and M. Hawkins, "Electrochemically Induced Nuclear Fusion of Deuterium," *Journal of Electroanalytical Chemistry* 261 (1989): 301, and erratum in *Journal of Electroanalytical Chemistry*, 263.

7. Gary Taubes, *Bad Science: The Short Life and Weird Times of Cold Fusion* (New York: Random House, 1993), 503.

8. G. Taubes, "Cold Fusion Conundrum at Texas A&M," *Science* 248 (1990): 1299.

9. D. M. Anderson and J. O. M. Bockris, "Cold Fusion at Texas A&M," *Science* 249 (1990): 463.

10. John R. Huizenga, *Cold Fusion: The Scientific Fiasco of the Century*, 2nd ed. (New York: Oxford University Press, 1993), 319.

11. J. Maddox, "End of Cold Fusion in Sight," *Nature* (London) 340 (1989): 15.

12. Frank Close, *Too Hot to Handle: The Race for Cold Fusion*, 2nd ed. (New York: Penguin, 1992).

13. Eugene F. Mallove, *Fire from Ice* (New York: John Wiley, 1991).

14. Nate Hoffman, *A Dialogue on Chemically Induced Nuclear Effects: A Guide for the Perplexed about Cold Fusion* (La Grange Park, Ill.: American Nuclear Society, 1995).

15. Hideo Kozima, *Discovery of the Cold Fusion Phenomenon: Development of Solid State Nuclear Physics and the Energy Crisis in the 21st Century* (Tokyo, Japan: Ohotake Shuppan Inc., 1998).

16. Tadahiko Mizuno, *Nuclear Transmutation: The Reality of Cold Fusion* (Concord, N.H.: Infinite Energy Press, 1998), 151.

17. Charles G. Beaudette, *Excess Heat: Why Cold Fusion Research Prevailed* (Concord, N.H.: Oak Grove Press, 2000).

18. Michael Shermer, *The Borderlands of Science: Where Sense Meets Nonsense* (Oxford, UK: Oxford University Press, 2001).

19. Bart Simon, *Undead Science: Science Studies and the Afterlife of Cold Fusion* (New Brunswick, N.J.: Rutgers University Press, 2002), 252.

20. Roberto Germano, *Fusione fredda: Moderna storia d'inquisizione e d'alchimia* (Napoli, Italy: Bibliopolis, 2003).

21. Steven B. Krivit and Nadine Winocur, *The Rebirth of Cold Fusion: Real Science, Real Hope, Real Energy* (Los Angeles: Pacific Oaks Press, 2004).

22. H. Kozima, *The Science of the Cold Fusion Phenomenon* (n.p.: Elsevier Science, 2006), 208.

23. Jed Rothwell, *Cold Fusion and the Future* (LENR-CANR.org 2007). The full text is also available as a PDF file at: www.LENR.org (accessed June 17, 2012). (not in print; available as a Kindle download).

24. E. Sheldon, "An Overview of Almost 20 Years' Research on Cold Fusion," *Contemporary Physics* (49) (5) (2009): 375.

25. D. Morrison, "A View from CERN," *Physics World* 2 (1989): 17.

26. Robert L. Park, *Voodoo Science* (New York: Oxford University Press, 2000).

27. See Park's website, What's New, at www.bobpark.org (accessed June 17, 2012).

28. N. S. Lewis, C. A. Barnes, M. J. Heben, et al., "Searches for Low-Temperature Nuclear Fusion of Deuterium in Palladium," *Nature* (London) (340) (6234) (1989): 525.

29. Energy Research Advisory Board, *Report of the Cold Fusion Panel to the Energy Research Advisory Board.* U.S. Department of Energy, DOE/S-0073 (Washington, D.C., 1989).

30. U.S. Department of Energy, *Report of the Review of Low Energy Nuclear Reactions,* in *Review of Low Energy Nuclear Reactions,* U.S. Department of Energy, Office of Science (Washington, D.C., 2004).

31. Peter L. Hagelstein, Michael C. H. McKubre, David J. Nagel, et al., *New Physical Effects in Metal Deuterides,* in *DOE Evaluation of Low Energy Nuclear Reactions,* U.S. Department of Energy (Washington, D.C., 2004). Also available as a PDF file at: www.LENR-CANR.org (accessed June 17, 2012).

32. I. Dardik, T. Zilov, H. Branover, et al., "Excess Heat in Electrolysis

Experiments at Energetics Technologies," in *11th International Conference on Cold Fusion* (Marseilles, France: World Scientific Co., 2004).

33. D. Clery, "Fusion Research: Design Changes Will Increase ITER Reactor's Cost," *Science* 320 (2008): 1405.

34. D. Clery, "Schedule Concerns Delay ITER's Go-Ahead," *Science* 326 (2009): 1172.

35. A. Karabut, "Research into Low-Energy Nuclear Reactions in Cathode Sample Solid with Production of Excess Heat, Stable and Radioactive Impurity Nuclides," in *Condensed Matter Nuclear Science, ICCF-12* (Yokohama, Japan: World Scientific, 2005).

36. T. N. Claytor, et al., "Tritium Production from Palladium Alloys," in *The Seventh International Conference on Cold Fusion, Vancouver, Canada, 1998* (Salt Lake City, Utah: ENECO Inc., 1998).

37. R. Stringham, "Low Mass 1.6 MHz Sonofusion Reactor," in *11th International Conference on Cold Fusion* (Marseilles, France: World Scientific Co., 2004).

38. S. J. Putterman, "Sonoluminescence: Sound into Light," *Scientific American* 272 (1995): 46.

39. R. P. Taleyarkhan, C. D. West, J. S. Cho, et al., "Evidence for Nuclear Emissions during Acoustic Cavation," *Science* 295 (2002): 1868.

40. S. B. Krivit, "Rusi Teleyarkhan Bubblegate Investigation Portal," *New Energy Times*, www.newenergytimes.com/v2/bubblegate/BubblegatePortal.shtml (accessed June 17, 2012).

41. Y. Arata and Y. C. Zhang, "Excess Heat and Mechanism in Cold Fusion Reaction," in *Fifth International Conference on Cold Fusion, Monte-Carlo, Monaco, 1995* (Sophia Antipolis Cedex, France: IMRA Europe, 1995).

42. Y. Arata and Y. C. Zhang, "Development of 'DS-Reactor' as the Practical Reactor of 'Cold Fusion" Based on the 'DS-Cell' with 'DS-Cathode,'" in *Condensed Matter Nuclear Science, ICCF-12* (Yokohama, Japan: World Scientific, 2005).

43. Y. Iwamura, T. Itoh, and M. Sakano, *Nuclide Transmutation Device and Nuclide Transmutation Method* (n.p.: Mitsubishi Heavy Industries, Ltd., 2002).

44. Y. Iwamura, M. Sakano, and T. Itoh, "Elemental Analysis of Pd Complexes: Effects of D_2 Gas Permeation," *Japanese Journal of Applied Physics A* (41) (7) (2002): 4642.

45. C. Louis Kervran, *Biological Transmutations* (Brooklyn, N.Y.: Swan House Publishing Co., 1972).

46. C. Louis Kervran, *Biological Transmutation* (Wappinger's Falls, N.Y.: Beekman Publishers Inc., 1980).

47. H. Komaki, "An Approach to the Probable Mechanism of the Non-Radioactive Biological Cold Fusion or So-Called Kervran Effect (Part 2)," in *Fourth International Conference on Cold Fusion, Lahaina, Maui, 1993* (Palo Alto, Calif.: Electric Power Research Institute, 1993).

48. V. I. Vysotskii, A. A. Kornilova, and I. I. Samoylenko, "Experimental Discovery of Phenomenon of Low-Energy Nuclear Transformation of Isotopes ($Mn^{55}=Fe^{57}$) in Growing Biological Cultures," in *Sixth International Conference on Cold Fusion, Progress in New Hydrogen Energy, Lake Toya, Hokkaido, Japan* (Tokyo, Japan: New Energy and Industrial Technology Development Organization, Tokyo Institute of Technology, 1996).

49. Vladimir I. Vysotskii and Alla A. Kornilova, *Nuclear Transmutation of Stable and Radioactive Isotopes in Biological Systems* (Hertforshire, UK: Motilal Books, 2010).

50. V. Vysotskii, et al. "Successful Experiments on Utilization of High-Activity Waste in the Process of Transmutation in Growing Associations of Microbiological Cultures," in *Tenth International Conference on Cold Fusion* (Cambridge, Mass.: 2003).

51. V. Bendkowsky, B. Butscher, J. Nipper, et al., "Novel Binding Mechanism for Ultra-long Range Molecules." Search on the PDF site: arXiv:0809.2961v1 (accessed August 16, 2012).

52. J. Wang and L. Holmlid, "Rydberg Matter Clusters of Hydrogen (H_2)N with Well-Defined Kinetic Energy Release Observed by Neutral Time-of-Flight," *Chemical Physics* 277 (2002): 201.

53. S. Badiei, P.U. Andersson, and L. Holmlid, "High-Energy Coulomb Explosions in Ultra-dense Deuterium: Time-of-Flight-Mass Spectrometry with Variable Energy and Flight Length," *International Journal of Mass Spectrometry* 282 (2009): 70.

54. Randell L. Mills, *The Grand Unified Theory of Classical Quantum Mechanics* (Ephrata, Pa.: Cadmus Professional Communications, 2006), 1450.

CHAPTER 12. ZERO POINT
ENERGY CAN POWER THE FUTURE

1. Daniel Lapedes, ed., *McGraw-Hill Dictionary of Physics and Mathematics* (New York: McGraw-Hill, 1978).

2. U.S. Department of Energy Office of Science, "Zero Point Energy," *Ask a Scientist*, www.newton.dep.anl.gov/askasci/phy00/phy00034.htm (accessed June 16, 2012).

3. Thomas Valone, *Zero Point Energy: The Fuel of the Future* (Beltsville, Md.: Integrity Research Institute, 2007).

4. Eric Davis, et al., "Review of Experimental Concepts for Studying the Quantum Vacuum Field," in *Proceedings of Space Technology and Applications International Forum,* ed. Mohamed S. El-Genk (College Park, Md.: American Institute of Physics, 1999).

5. Robert Forward, "Extracting Electrical Energy from the Vacuum by Cohesion of Charged Foliated Conductors," *Physical Review B* (30) (4) (1984): 1700–1702.

6. Henry Bortman, "Energy Unlimited," *New Scientist* (165) (2222) (2000): 32.

7. Marc-Thierry Jaekel and Serge Reynaud, "Movement and Fluctuations of the Vacuum," *Reports on Progress in Physics* 60 (1997): 867.

8. Ibid., 879–80.

9. Margaret Hawton, "One-Photon Operators and the Role of Vacuum Fluctuations in the Casimir Force," *Physical Review A* (50) (2) (1994): 1057.

10. H. E. Puthoff, "Ground State of Hydrogen as a Zero-Point-Fluctuation-Determined State," *Physical Review D* (35) (10) (1987): 3266–3269.

11. H. E. Puthoff, "Why Atoms Don't Collapse," *New Scientist,* July 28, 1990.

12. Robert Forward, "An Introductory Tutorial on the Quantum Mechanical Zero Temperature Electromagnetic Fluctuations of the Vacuum," Mass Modification Experiment Definition Study. Phillips Laboratory Report PL-TR 96-3004 (1996).

13. S. K. Lamoreaux, *Physical Review Letters* 78 (n.d.): 1, 97.

14. US Patent 4,704,622. View any U.S. patent at www.uspto.gov (accessed June 16, 2012).

15. *Washington Times,* May 20, 2001, A3.

16. Bortman, "Energy Unlimited.

17. Chris Binns, "What Lies beneath the Void," University of Leicester, http://ebulletin.le.ac.uk/features/2000-2009/2005/08/nparticle-82w-fqr-2cd (accessed June 16, 2012). Binns, a professor of physics and astronomy, discusses a project connected to the "zero-point energy" of space.

18. Jim Wilson, Louis Brill, Stefano Coledan, et al., "Power from a Seething Vacuum," *Popular Mechanics* (179) (1) (January 2002): 22.

19. F. Pinto, "Engine Cycle of an Optically Controlled Vacuum Energy Transducer," *Physical Review B* (60) (21) (1999): 14, 740.

20. Fabrizio Pinto, "Progress in Quantum Vacuum Engineering: Nanotechnology and Propulsion" (keynote address, Second International Conference

on Future Energy, Washington D.C., September 23, 2006). DVD available from: Integrity Research Institute, 5020 Sunnyside Ave., Suite 209, Beltsville, Md., 20705.

21. Ibid.

22. Fritjof Capra, *The Tao of Physics* (Boston: Shambhala, 1975), 232, 304.

23. Davide Iannuzzi, Vrije University, Amsterdam, Netherlands, 2006, www .inrim.it/events/docs/Casimir_2006/Iannuzzi.pdf (accessed August 17, 2012).

24. J. Smoliner, W. Demmerle, E. Gornik, et al., "Tunelling Spectroscopy of 0D States," *Semiconductor Science and Technology* 9 (1994): 1925.

25. Thomas Valone, *Practical Conversion of Zero-Point Energy: Feasibility Study of the Extraction of Zero-Point Energy Extraction from the Quantum Vacuum for the Performance of Useful Work* (Beltsville, Md.: Integrity Research Institute, 2003).

26. Roger H. Koch, D. J. Van Harlingen, and John Clarke. "Measurements of Quantum Noise in Resistively Shunted Josephson Junctions," *Physical Review B* (26) (1) (July 1982): 74. Also see: R. H. Koch, D. J. Van Harlingen, and J. Clarke, "Quantum-Noise Theory for the Resistively Shunted Josephson Junction," *Physical Review Letters* 45 (1980): 2132–35; and: R. H. Koch, D. J. Van Harlingen, and J. Clarke, "Observation of Zero-Point Fluctuations in a Resistively Shunted Josephson Tunnel Junction," *Physical Review Letters* 47 (1981): 1216–19.

27. Davis, "Review of Experimental Concepts."

28. Christian Beck and Michael Mackey, "Could Dark Energy Be Measured in the Lab?" Preprint abstract. Cornell University Library, Astrophysics, June 23, 2004, www.arxiv.org/abs/astro-ph/0406504 (accessed June 16, 2012).

29. Jonathan Lynch, et al., "Unamplified Direct Detection Sensor for Passive Millimeter Wave Imaging," in *Passive Millimeter-Wave Imaging Technology IX,* ed. Roger Appleby, *Proceedings of SPIE* 6211 (2006): 621101; also see: J. N. Schulman and D. H. Chow, "Sb-Heterostructure Interband Backward Diodes," *Electron Device Letters, IEEE* 21 (2000): 353–55.

30. A. C. Young, J. D. Zimmerman, E. R. Brown, and A. C. Gossard, "Semi-metal-Semiconductor Rectifiers for Sensitive Room-Temperature Microwave Detectors," *Applied Physics Letters* 87 (2005).

31. Henrik Brenning, et al., *Journal of Applied Physics* 100 (2006): 114321.

32. C. Van den Broeck, "Brownian Refrigerator," *Physical Review Letters* 96 (2006): 210601.

33. Philip Ball, *Nature,* February 2004.

34. A. Feigel, "Quantum Vacuum Contribution to the Momentum of Dielectric Media," *Physical Review Letters* 92 (2004), doi:10.1103/PhysRevLett.92 .020404.

35. Ibid.

36. E. Iacopini, "Casimir Effect at Macroscopic Distances," *Physical Review A* (48) (1) (1993): 129.

37. S. Haroche and J. Raimond, "Cavity Quantum Electrodynamics," *Scientific American,* April 1993: 56.

38. Stefan Weigert, "Spatial Squeezing of the Vacuum and the Casimir Effect," *Physics Letters A* 214 (1996): 215.

39. Astrid Lambrecht, Marc-Thierry Jaekel, and Serge Reynaud, "The Casimir Force for Passive Mirrors," *Physics Letters A* 225 (1997): 193.

40. M. V. Cougo-Pinto, "Bosonic Casimir Effect in External Magnetic Field," *Journal of Physics A: Mathematical and General* 32 (1999): 4457.

41. Forward, "Introductory Tutuorial," 3.

42. Pinto, "Progress in Quantum Vacuum Engineering."

43. Ilya Prigogine, *Order out of Chaos: Man's New Dialogue with Nature* (New York: Bantam Books, 1984).

44. Ibid., 286.

45. Ibid., 276 (Prigogine references two of his own works here: M. Courbage and I. Prigogine, "Intrinsic Randomness and Intrinsic Irreversibility in Classical Dynamical Systems," *Proceedings of the National Academy of Sciences* 80 (April 1983); and I. Prigogine and Cl. George, "The Second Law as a Selection Principle: The Microscopic Theory of Dissipative Processes in Quantum Systems," *Proceedings of the National Academy of Sciences* 80 (April 1983).

46. Prigogine, *Order Out of Chaos,* 181.

47. Kevin Bullis, "Tiny Solar Cells," *Technology Review,* October 18, 2007.

AFTERWORD

1. James Petras, "The Ecuadorian Coup: Its Larger Meaning," *Centre for Research on Globalization,* www.globalresearch.ca/index.php?context =va&aid=21377 (accessed June 17, 2012). This is an excellent review of Ecuador's politics, written October 2010; also the websites for Amazon Watch (www.amazonwatch.org, accessed June 17, 2012) and Upside Down World (www.upside downworld.org, accessed June 17, 2012) have informative postings of the politics of the rain forest.

2. The legal case for reparations from the Chevron-Texaco Ecuador oil spill is well described in the 2009 film *Crude*.

3. Matt Finer, Clinton N. Jenkins, Stuart L. Pimm, et al., "Oil and Gas Projects in the Western Amazon: Threats to Wilderness, Biodiversity, and Indigenous Peoples," *Save America's Forests*, www.saveamericasforests.org/WesternAmazon/index.html (accessed June 17, 2012).

4. There are many summaries and status reports on the Yasuni initiative to keep the oil in the ground, for example, in the 2010 film *Yasuni—Two Seconds of Life*, www.yasuni-film.com (accessed June 17, 2012).

5. Brian O'Leary, "The Ecuador Initiative: How Innovation Could Save the Amazon Rainforest and Other Habitats While Creating Economic Sovereignty for Ecuador," *Dr. Brian O'Leary: Blog*, www.drbrianoleary.wordpress.com (accessed June 17, 2012).

ADDITIONAL RESOURCES

INTRODUCTION—SUPPLEMENTAL RESOURCES

Beaudette, Charles G. *Excess Heat: Why Cold Fusion Research Prevailed.* Oak Grove Press, 2002.

Krivit, Steven B., and Nadine Winocur. *The Rebirth of Cold Fusion.* Pacific Oaks Press, 2004.

Mallov, Eugene. *Fire from Ice: Searching for the Truth behind the Cold Fusion Furor.* Infinite Energy Press, 1999.

Storms, Edmund. *Science of Low Energy Nuclear Reaction: A Comprehensive Compilation of Evidence and Explanations about Cold Fusion.* World Scientific Publishing Co., 2007.

Eugene F. Mallove, Part 1, www.youtube.com/watch?v=O1FSpky-JXg (accessed June 18, 2012).

Eugene F. Mallove, Part 2, www.youtube.com/watch?v=I2W9A7ufAdE&feature=relmfu (accessed June 18, 2012).

Heavy Watergate: The War on Cold Fusion, www.youtube.com/watch?v=htgV7fNO-2k (accessed June 18, 2012).

Jhon, Mu Shik. *The Water Puzzle and the Hexagonal Key.* Trans. and ed. M. J. Pangman. Uplifting Press Inc., 2004.

Juhasz, Antonia. *The Tyranny of Oil.* William Morrow, 2008.

Kuhn, Thomas S. *The Structure of Scientific Revolutions,* 3rd ed. University of Chicago Press, 1996.

Laszlo, Ervin. *You Can Change the World.* Select Books Inc., 2003.

———. *Science and the Akashic Field.* Inner Traditions, 2004.

LaViolette, Paul A., Ph.D. *Secrets of Antigravity Propulsion: Tesla, UFOs, and Classified Aerospace Technology,* 3rd ed. Bear & Company, VT, 2008.

———. *Subquantum Kinetics: A Systems Approach to Physics and Cosmology,* 2nd ed. Starline Publications, 2008.

Mitchell, Edgar. *The Way of the Explorer.* Career Press Inc., 2008.

Moray, T. Henry. *The Sea of Energy.* Cosray Research Institute, 1978. Reprint; original title, *The Sea of Energy in Which the Earth Floats.*

Ray, Paul H., and Sherry Ruth Anderson. *The Cultural Creative.* Three Rivers Press, 2000.

Sheldrake, Rupert. *Seven Experiments That Could Change the World: A Do-It-Yourself Guide to Revolutionary Science,* 2nd ed. Park Street Press, 2002.

Sigma, Rho, and Edgar Mitchell. *Ether-Technology: A Rational Approach to Gravity Control.* Adventures Unlimited Press, 1996.

Tiller, William A., Walter E. Dibble Jr., and Michael J. Kohane. *Conscious Act of Creation: The Emergence of a New Physics.* Pavior, 2001.

Tutt, Keith. *The Scientist, the Madman, the Thief, and Their Lightbulb.* Simon and Schuster UK Ltd., 2001.

Valone, Thomas. *Bioelectromagnetic Healing: A Rationale for Its Use.* Integrity Research Institute, 2000.

Vassilatos, Gerry. *Lost Science.* Adventures Unlimited Press, 1999.

Websites

John Bedini, www.johnbedini.net/john34/bedinibearden.html (accessed June 18, 2012).

Institute of Noetic Sciences, www.noetic.org (accessed June 18, 2012).

Pure Energy Systems: PESWiki, "New Energy Congress," www.peswiki.com/index.php/New_Energy_Congress (accessed June 18, 2012).

A library of papers on low-energy nuclear reactions (LENR), also known as cold fusion and chemically assisted nuclear reactions (CANR), is available at www.lenr-canr.org (accessed June 27, 2012).

YouTube Videos

Energy from the Vacuum, documentary series trailer, www.youtube.com/watch?v=X6EnDBjCjBw (accessed June 18, 2012).

Energy from the Vacuum: Part 6—Inside Radiant Energy, John Bedini. A docu-

mentary series. Part 6 features dialogues with John Bedini. www.youtube
.com/watch?v=0PBHePZ_6U8 (accessed August 16, 2012).

ZPE lecture, www. youtube.com/watch?v=OlgSiUJthco

DVDs

Burn Up. Exposé of the oil industry starring Neve Campbell and Bradley
Whitford. PAL region 2.4 import from the United Kingdom. Occasionally
aired on Planet Green TV.

The 11th Hour, narrated by Leonardo DiCaprio. Warner Video.

Energy from the Vacuum: Part 2—John Bedini. A documentary series. www.che
niere.org/sales/buy-e2.htm (accessed August 16, 2012).

An Inconvenient Truth, with Al Gore. Paramount Home Entertainment.

The Sea of Energy, with John E. Moray. Video on the work of T. Henry Moray.
House of Moray.

CHAPTER 1. NIKOLA TESLA

Books and Periodicals

O'Neal, John J. *Prodigal Genius: The Life of Nikola Tesla.* Adventures Unlimited
Press, 2008.

Seifer, Marc J. *Wizard: The Life and Times of Nikola Tesla.* Carol Publishing
Group, 1999.

Tesla, Nikola. *The Fantastic Inventions of Nikola Tesla.* Adventures Unlimited
Press, 1993.

———. *My Inventions: The Autobiography of Nikola Tesla.* CreateSpace,
2010.

Websites

Charged: The Story of Nikola Tesla, www.theteslamovie.com (accessed June 17,
2012).

Marc Seifer, www.marcseifer.com (accessed June 17, 2012).

The Turn of the Century Electrotherapy Museum, www.electrotherapymuseum.com
(accessed June 17, 2012).

YouTube Videos

Nikola Tesla: The Lost Wizard (short demonstration film for proposed full-
length "biopic" on Tesla). New York City Mayor Fiorello LaGuardia,
in a stirring voice, reads a eulogy upon the death of Nikola Tesla, slides

chosen by Marc J. Seifer, edited by Tim Eaton, www.youtube.com/watch?v=MldwLN5WjwY (accessed July 3, 2012).

Nikola Tesla: The Missing Secrets, part 1, www.youtube.com/watch?v=ZT64IzfRkK4&feature=relmfu (accessed June 17, 2012).

Nikola Tesla: The Missing Secrets, part 2, www.youtube.com/watch?v=RKbXfksgNEA&feature=relmfu (accessed June 17, 2012).

Nikola Tesla: The Missing Secrets, part 3, www.youtube.com/watch?v=Tfx6vSJELF0&feature=relmfu (accessed June 17, 2012).

Nikola Tesla: The Missing Secrets, part 4, www.youtube.com/watch?v=TaX_HYi0Omc&feature=relmfu (accessed June 17, 2012).

Nikola Tesla: The Missing Secrets, part 5, www.youtube.com/watch?v=ohbseq7Nez4&feature=relmfu (accessed June 17, 2012).

Documentary Short Films

Two short documentaries on Tesla's life based on Marc Seifer's biography and screenplay for the proposed film *Tesla: The Lost Wizard,* cowritten with visual effects editor Tim Eaton, formerly with Lucasfilms ILM and Sony Imageworks. www.youtube.com/watch?v=T2PyyO1nv7I (accessed August 13, 2012).

DVDs

Nikola Tesla: The Genius Who Lit the World. English, all regions. Directors: Dr. Ljubo Vujovic, Professor Aleksandar Marincic, Henry Jesionka

Tesla: Master of Lighting. English, region 1. PBS Home Video.

CHAPTER 2.
JOHN WORRELL KEELY

Books and Periodicals

Blavatsky, Helena Petrovna, *The Secret Doctrine,* vol. I. Theosophical University Press, 1970, 554–65.

Bloomfield-Moore, Clara. *Keely's Secrets.* T.P.S., 1888.

———. *Keely and His Inventions: Aerial Navigation.* Kegan Paul, Trench, Trubner and Co., 1893.

———. *Keely and His Discoveries,* 2nd ed. Kessinger Publishing, 2004.

Burridge, Gaston. "The Baffling Keely Free Energy Machines." *Fate* 10, no. 7 (1957).

Crossen, Joseph B. *Keely: Quack or Visionary?* The Maple Press Company, 1972.

Edwards, Frank. *Strangest of All*. Citadel, 1956.

Fort, Charles. *Wild Talents*. Claude Kendall, 1932.

Paijmans, Theo. *Free Energy Pioneer: John Worrell Keely*. Adventures Unlimited Press, 2004.

Pond, Dale. *The Physics of Love: The Ultimate Universal Laws*. The Message Company, 1996.

——. *Universal Laws Never before Revealed: Keely's Secrets; Understanding and Using the Science of Sympathetic Vibration*, rev. ed. The Message Company, 2007.

Sykes, Egerton. *The Keely Mystery*. Markham House, 1972, available from www .svpbookstore.com (accessed June 17, 2012).

——. *The Keely Photographs*. Brighton, 1973.

Wendelholh AB, G. *Keely: Pictures of His Discoveries*. G. Wendelholh AB, n.p., 1987.

Websites

Davidson, Dan A. "Free Energy, Gravity, and the Aether," www.keelynet.com/ davidson/npap1.htm (accessed June 17, 2012).

Jessup and Moore: "Poetry and Paper Making," www.holtermann.se/jessup_moore/ jesmore/index.html (accessed June 17, 2012).

Pure Energy Systems: PESWiki, "PowerPedia: John Keely," www.peswiki.com/ index.php/PowerPedia: John_Keely (accessed June 17, 2012).

Sympathetic Vibratory Physics: A Musical Universe, www.svpvril.com (accessed June 17, 2012).

Videos

Pond, Dale, *Dale Pond: The Basic Principles of SVP 1/2— John Worrell Keely* (ninety-minute lecture), www.video.google.com/videoplay? docid=9125003792513982191 (accessed June 17, 2012).

CHAPTER 3.
VICTOR SCHAUBERGER

Books and Periodicals

Alexanderson, Olof. *Living Water: Victor Schauberger and the Secrets of Natural Energy*. Gill and Macmillan Ltd, 1990. (This was the first book in English about Viktor Schauberger.)

Batmanghelidj, Fereydoon, *Your Body's Many Cries for Water*. Global Health Solutions Inc., 2010.

Coats, Callum. *Living Energy*. Gateway, 2002.

Cobbald, Jane. *Viktor Schauberger: A Life of Learning from Nature*. Floris Books, 2009.

Emoto, Masaru. *The Secret Life of Water*. Atria Books, 2005.

———. *The True Power of Water*. Atria Books, 2005.

———. *Messages from Water and the Universe*. Hay House, 2010.

Schauberger, Viktor. The Ecotechnology Series, trans. and ed. Callum Coats. (Schauberger's writings, encompassing four volumes: *The Water Wizard*, Gateway, 1998; *Nature as Teacher*, Gateway, 1998; *The Fertile Earth*, Gateway, 2001; and *The Energy Evolution*, Gateway, 2001.)

Schwenk, Theodor. *Sensitive Chaos*. Steiner, 1965.

Steiner, Rudolf. *The Nature of Substance*. Steiner, London, 1966.

YouTube Videos

Nature Was My Teacher: The Vision of Viktor Schauberger, www.youtube.com/watch?v=3wl-Temag9E&feature=related (accessed June 17, 2012).

DVDs

The Extraordinary Nature of Water, http://video.google.com/videoplay?docid=-8915966819502040048 (accessed August 13, 2012)

Water—The Great Mystery, www.youtube.com/watch?v=6Xfv2-riA2g (accessed August 13, 2012)

CHAPTER 4.
ROYAL RAYMOND RIFE

Books and Periodicals

Foye, Gerald F. *Royal R. Rife: Humanitarian, Betrayed and Persecuted*. New Century Press, 2002.

Lynes, Barry. *The Cancer Cure that Worked: 50 Years of Suppression*. Compcare Publications, 1987.

Vassilatos, Gerry. *Lost Science*. www.borderlands.com (accessed August 22, 2012).

YouTube Videos

Royal Raymond Rife: Suppressed Medical Technology, www.youtube.com/watch?v=2fh0RJczTAc (accessed June 18, 2012).

Royal Rife—In His Own Words, www.youtube.com/watch?v=W4JtNQnnut8& feature=related (accessed June 18, 2012).

DVDs

The Rise and Fall of a Scientific Genius: The Forgotten Story of Royal Raymond Rife. Producer: Sean Montgomery, 2007. Available at www.zeroz erotwo.org (accessed August 17, 2012).

CHAPTER 5.
T. TOWNSEND BROWN

Books and Periodicals

Brown, Paul. "Electrostatic Propulsion References." *Electric Spacecraft Journal,* January–March 1992.

Burridge, Gaston. "Townsend Brown and His Anti-gravity Discs." *Fate,* November 1958.

Hall, Steve. "The Electrokinetic Works of T. T. Brown." *Electric Spacecraft Journal,* January–March 1991.

LaViolette, Paul, Ph.D. "Electrogravitics: Back to the Future." *Electric Spacecraft Journal,* October–December 1991.

———. "The U.S. Antigravity Squadron." In *Proceedings of the International Symposium on New Energy* (Denver, Colo.: Rocky Mountain Research Institute, 1993).

———. *Secrets of Antigravity Propulsion: Tesla, UFOs, and Classified Aerospace Technology,* 3rd ed. Bear & Company, 2008 (contains updated information on T. T. Brown's inventions).

Millis, Marc, and Eric Davis. *Frontiers of Propulsion Science.* AIAA, 2009.

Moore, William L., and Charles Berlitz. *The Philadelphia Experiment: Project Invisibility.* Ballantine Books, 1979.

Nichelson, Oliver. Letter to the editor. *Electric Spacecraft Journal,* April–June 1992.

Richards, R. Louis. "Resonant Vortex." *Borderlands,* third quarter, 1993.

Schaffranke, Rolf, Ph.D. *Ether Technology: A Rational Approach to Gravity Control.* Cadake Publishing, 1977.

Yost, Charles A. "T. T. Brown and the Bahnson Lab Experiments." *Electric Spacecraft Journal,* April–June 1991.

———. "Electric Propulsion Research." *Electric Spacecraft Journal,* July–September 1991.

CHAPTER 6. THE SUSTAINABLE
TECHNOLOGY SOLUTION REVOLUTION

Books and Periodicals

Aronson, James, Suzanne J. Milton, James N. Blignaut, and Peter H. Raven. *Restoring Natural Capital: Science, Business, and Practice*. Island Press, 2007.

Bartholomew, Alick. *The Story of Water: Source of Life*. Floris Books, 2010.

Dickinson, Joel, with Robert Cook. *The Death of Rocketry*. CPI Systems Inc., 1980.

O'Leary, Brian. *Exploring Inner and Outer Space: A Scientist's Perspective on Personal and Planetary Transformation*. North Atlantic Books, 1989.

———. *The Second Coming of Science: An Intimate Report on the New Science*. North Atlantic Books, 1992.

———. *Miracle in the Void*. Kanapua'a Press, 1996.

———. *Re-inheriting the Earth: Awakening to Sustainable Solutions and Greater Truths*. Truth Seeker, 2003.

———. *The Energy Solution Revolution*, 2nd ed. Bridger House, 2009.

Pauli, Gunter. *The Blue Economy: 10 Years, 100 Innovations, 100 Million Jobs*. Paradigm Publications, 2010.

Perkins, John. *Hoodwinked*. Broadway Books, 2009.

Quinn, Daniel. *Ishmael: An Adventure of the Mind and Spirit*. Bantam, 1995.

Yurth, David. *The Ho Chi Minh Handbook of Guerilla Warfare: Strategies for Innovation Management*. In press, 2010.

Websites

Kelly, Patrick. "Practical Guide to Free Energy Devices." www.free-energy-info .com (accessed June 18, 2012). This is an excellent review of the state of the art of new energy technologies, including descriptions of specific devices.

O'Leary, Brian. "The Ecuador Initiative: How Innovation Could Save the Amazon Rainforest and Other Habitats While Creating Economic Sovereignty for Ecuador." *Dr. Brian O'Leary: Blog,* www.drbrianoleary. wordpress.com (accessed June 18, 2012).

Society for Ecological Restoration International, www.ser.org (accessed June 18, 2012)

Vesperman, Gary. "History of 'New Energy' Invention Suppression Cases." www.rense.com/general72/oinvent.htm (accessed June 18, 2012).

Zero Emissions Research and Initiatives, www.zeri.org (accessed June 18, 2012). Contains information about Gunter Pauli and the blue economy.

Miscellaneous

Infinite Energy, the magazine of new energy science and technology. Subscriptions and more information at: www.infinite-energy.com (accessed June 18, 2012).

Foundations

New Energy Foundation Inc.
P.O. Box 2814
Concord, New Hampshire 03302–2816

Videos

Yasuni—Two Seconds of Life. Producer: Eric Spitzer. www.yasuni-film.com (accessed June 18, 2012).

CHAPTER 7. POWER FOR THE PEOPLE— FROM WATER

Books and Periodicals

Manning, Jeane, and Joel Garbon. *Breakthrough Power: How Quantum-Leap New Energy Inventions Can Transform Our World.* Amber Bridge Publishing, 2009. For more information, see www.breakthroughpower.net (accessed June 18, 2012).

Robey, James. *Water Car: How to Turn Water into Fuel.* Kentucky Water Fuel Museum, 2006.

Schauberger, Viktor, *The Energy Evolution,* trans. and ed. Callum Coats. Gateway, 2001.

Websites

BlackLight Power, www.blacklightpower.com (accessed June 18, 2012).

Changing Power (Jeane Manning), www.jeanemanning.com (accessed June 18, 2012).

Eagle-Research, www.eagle-research.com (accessed June 18, 2012).

New Energy Movement, Jeane Manning and Joel Garbon interview about breakthrough energy on "Free Energy Now" radio, www.newenergymovement .org/radioi.php (accessed June 18, 2012).

Pure Energy Systems: PESWiki, "Directory: Hydroxy or HHO Injection Systems," www.peswiki.com/index.php/Directory:Hydroxy_or_HHO_Injection_Systems (accessed June 18, 2012).

Pure Energy Systems: PESWiki, "New Energy Congress," www.peswiki.com/index.php/New_Energy_Congress (accessed June 18, 2012).

CHAPTER 8. IMAGINE A FREE ENERGY FUTURE FOR ALL OF MANKIND

Books and Periodicals

LaViolette, Paul. *Secrets of Antigravity Propulsion: Tesla, UFOs, and Classified Aerospace Technology,* 3rd ed. Bear & Company, 2008.

Valone, Thomas, ed. *Electrogravitics Systems: Reports on a New Propulsion Methodology.* Integrity Research Institute, 1993.

———. *Electro Gravitics II: Validating Reports on a New Propulsion Methodology,* 2nd ed. Integrity Research Institute, 2005.

Websites

LaViolette, Paul, "Moving beyond the First Law and Advanced Field Propulsion Technologies," *The Orion Project,* www.theorionproject.org/en/movingbeyond_laviolette.html (accessed June 18, 2012).

Naudin, Jean-Louis (JLN Labs), "The Quest for Overunity," www.jlnlab.com (accessed June 18, 2012).

The Orion Project, www.theorionproject.org (accessed June 18, 2012).

Sphinx Stargate, information on Paul LaViolette, www.etheric.com (accessed June 18, 2012).

Thomas Bearden, Website on energy technologies and issues, www.cheniere.org (accessed June 18, 2012).

CHAPTER 10. HARNESSING NATURE'S FREE ENERGY

Books and Periodicals

LaViolette, Paul A. "The Searl Effect." In *Secrets of Antigravity Propulsion: Tesla, UFOs, and Classified Aerospace Technology,* 3rd ed. Bear & Company, 2008, 296–332.

Thomas, John A., Jr., *Antigravity: The Dream Made Reality; The Story of John R. R. Searl.* DISC Worldwide Inc., 1993. (Available at www.searleffect.com, accessed July 3, 2012.)

Websites

The Searl Solution, www.searlsolution.com (accessed June 18, 2012).

YouTube Videos

SEG Mockup Scope DVD (Free Energy Generator), www.youtube.com/watch?v=w1tYhCFksqE (accessed June 18, 2012).

DVDs

The John Searle Story: Keepers of the Earth, 2008. Documentary film, www.johnsearlstory.com/about.html (accessed August 16, 2012).

CHAPTER 11. COLD FUSION

Books and Periodicals

Aspden, Harold. *Modern Aether Science.* Sabberton Publications, 1972.

Bourne, Edmund J. *Global Shift.* New Harbinger Publications Inc., 2008.

Cook, Nick. *The Hunt for Zero Point.* Broadway Books, 2001.

Eisen, Jonathan. *Suppressed Inventions and Other Discoveries.* Perigee, 1999.

Harmon, Willis, and Jane Clark. *New Metaphysical Foundations of Modern Science.* Institute of Noetic Sciences, 1994.

Hellyer, Paul. *Light at the End of the Tunnel: A Survival Plan for the Human Species.* AuthorHouse, 2010.

CHAPTER 12. ZERO POINT ENERGY CAN POWER THE FUTURE

Books and Periodicals

A number of popular books introduce zero point energy with very readable details. For example, the recent book *The Fabric of the Cosmos* by Brian Greene has a short section called "Quantum Jitters and Empty Space" with good graphics and a discussion of the Casimir effect (Vintage Books, 2004, 329–35). Also, *Einstein for Dummies* by Carlos Calle has a short section on zero point energy and the Casimir force (Wiley Publishing, 2005, 312).

King, Moray B. *Quest for Zero Point Energy Engineering Principles for Free Energy.* Adventures Unlimited Press, 2002.

———. *Tapping the Zero-Point Energy.* Adventures Unlimited Press, 2002.

Valone, Thomas. *Practical Conversion of Zero-Point Energy: Feasibility Study of the Extraction of Zero-Point Energy Extraction from the Quantum Vacuum for the Performance of Useful Work.* Integrity Research Institute, 2005.

————. *Zero Point Energy: The Fuel of the Future.* Integrity Research Institute, 2007.

DVDs

Free Energy and Antigravity Propulsion: Miracle in the Void. Featuring Dr. Tom Valone and Dr. Brian O'Leary. Two-DVD special edition, English, all regions.

CONTRIBUTORS

Callum Coats is the author of *Living Energies*, a detailed overview of the environmental theories of Austrian forester and natural scientist Viktor Schauberger (1885–1958), and translator-cum-editor of a representative selection of Schauberger's writings encompassed in the four books of the Ecotechnology Series, namely *The Water Wizard, Nature as Teacher, The Fertile Earth*, and *The Energy Evolution*. Coats resides in Queensland, Australia.

Tom Engelhardt is the cofounder of the American Empire Project. His books include *The American Way of War: How Bush's Wars Became Obama's* as well as *The End of Victory Culture: Cold War America and the Disillusioning of a Generation*, and he runs the Nation Institute's TomDispatch.com website. His latest book is *The United States of Fear* (Haymarket Books).

Finley Eversole, Ph.D., is a philosopher, educator, activist, and advocate for the role of the arts in the evolution of consciousness. In the 1960s he was active in the civil rights and women's movements and participated in organizing the first Earth Day in New York City in 1970. As executive director of the Society for the Arts, Religion, and Contemporary Culture, he worked with such cultural leaders as Joseph Campbell, W. H. Auden, Alan Watts, and Marianne Moore. Eversole edited and contributed to the book *Christian Faith and the Contemporary Arts* and is the author of *Art and Spiritual Transformation*. He has planned and edited five forthcoming volumes addressing solutions to a range of global problems; *Infinite Energy Technologies* is volume one in this series. Eversole is biographed in Marquis *Who's Who in the World 2011–2013* as well as the (thirty-year) *Pearl Anniversary Edition*.

Steven M. Greer, M.D., is the founder and director of the Disclosure Project. A lifetime member of Alpha Omega Alpha, the nation's most prestigious medical honor society, Greer has retired from his position as an emergency room physician to work pursuing a worldwide search for alternative energy sources through the group he founded: the Orion Project. Specifically, the Orion Project is seeking to bring to the public the energy sources known as zero point energy (or overunity energy), with the plan to identify and develop systems that will eliminate the need for fossil fuels. Greer has appeared on CNN, *Larry King Live Special,* the BBC, CBS, NHK (Japan), Telemundo, and numerous other TV and radio shows around the world, and he was a speaker at the 2011 Sundance Film Festival.

Theodore C. Loder III, Ph.D., received his doctorate from the University of Alaska in chemical oceanography, his postdoctorate degree from Dalhousie University, and served as a full professor at the University of New Hampshire from 1972 to 2005. In 2001, he coauthored the briefing document for the Disclosure Project. In 2002, as a member of the Space Colonization Technical Committee of the American Institute of Aeronautics and Astronautics, he wrote the included paper on energy technologies for the twenty-first century. He has given many lectures on both UFOs and new energy technologies and is presently on the boards of the Orion Project, the Center for the Study of Extraterrestrial Intelligence, and the New Energy Foundation.

Jeane Manning has traveled in twelve countries and interviewed dozens of scientists since 1981, researching revolutionary clean energy systems that could replace oil. With Joel Garbon, she coauthored the award-winning book *Breakthrough Power: How Quantum-Leap New Energy Inventions Can Transform Our World*. Her earlier books include *The*

Coming Energy Revolution and *Energie,* and several coauthored books, including *Angels Don't Play This HAARP* with Dr. Nick Begich. Her books have been published in seven languages. Manning now lives near Vancouver, Canada. Her websites are www.BreakthroughPower.net and www.ChangingPower.net.

Brian O'Leary, Ph.D., who passed away in July 2011, was a scientist-philosopher with fifty years of experience in academic research, teaching, and government service in frontier science and energy policy. A NASA scientist-astronaut during the Apollo program, he was the first to be selected for a planned Mars mission. Over the past four decades, O'Leary was an international author, speaker, peace activist, founder of nonprofits, and advisor to progressive U.S. Congress members and presidential candidates. His book, *The Energy Solution Revolution*, describes the enormous potential of breakthrough clean energy technologies, their suppression, and their logical necessity for our survival. In 2004, he and his wife, artist Meredith Miller, moved to the Andes in Ecuador, where they cocreated Montesueños, an eco-retreat and educational center dedicated to creativity and the rights of nature.

Theo Paijmans is the author of *Free Energy Pioneer: John Worrell Keely*. Published in 1998, it has been reprinted and a translated version was published in Japan in 2000. His articles and papers have appeared in various publications including *All Hallows, Strange Attractor, The Anomalist, Gazette Fortéenne,* CFZ Yearbooks 2009 and 2010, several *Darklore* volumes, and *Fortean Times,* where he is a regular columnist. He appeared as an expert in *Dark Fellowships: The Vril Society,* the Discovery Channel documentary on the Vril Society, which is the subject of his upcoming book. He is always interested in learning more. For this, seriously minded parties may contact him at th.paijmans@wxs.nl.

John L. Petersen is president and founder of the Arlington Institute, a nonprofit, future-oriented research institute. He is best known for writing and thinking about high impact surprises—wild cards—and the process of surprise anticipation. He was a naval flight officer in the U.S. Navy and Navy Reserve and is a decorated veteran of both the Vietnam and Persian Gulf Wars. He has served in senior positions for a number of presidential political campaigns and was an elected delegate to the Democratic National Convention in 1984. His books include *The Road to 2015, Out of the Blue: How to Anticipate Wild Cards, Big Future Surprises,* and *A Vision for 2012: Planning for Extraordinary Change.* His website is www.arlingtoninstitute.org.

John R. R. Searl invented and built the first Searl Effect Generator (SEG) in 1946. In 1990 he met John A. Thomas Jr. and began collaborating with him. In 1993, he released book one of *The Law of the Squares,* published by Thomas. Between 1993 and 2010 he wrote twenty-four additional books, also published by Thomas. In 1996, he founded many different companies and websites related to his technological innovations. More recently he has lectured widely around the world and directed the building of the SEG website: www.searlsolution .com. He has been listed in *Who's Who in the World* from 1993 on, and in addition to his many books published, is the author of numerous newspaper articles. He lives in the United Kingdom.

Marc J. Seifer, Ph.D., is the author of *Wizard: The Life and Times of Nikola Tesla* (Citadel Press). Another book by Seifer, *Transcending the Speed of Light* (Inner Traditions), discusses Nikola Tesla's theories on cosmology, gravity, ether, and the God particle, and their comparison to Einstein's theory of relativity on the fundamental structure of spacetime. In addition to appearances on the History Channel and *Coast-to-Coast AM* radio, he has lectured on Tesla at West Point Military Academy; the New York Public Library; IEEE meetings in Toronto and Colorado Springs; international conferences in Belgrade, Serbia, and Zagreb, and Croatia; LucasFilms Industrial Light and Magic; and the United Nations. He teaches psychology at Roger Williams University in Bristol, Rhode Island. For more information please visit: www.MarcSeifer.com.

Edmund Storms, Ph.D., received his doctorate in radiochemistry from Washington University in St. Louis and retired from the Los Alamos National Laboratory after thirty-four years of service. He presently lives in Santa Fe, where he is investigating the cold fusion effect in his own laboratory. His book on the subject, *The Science of Low Energy Nuclear Reaction: A Comprehensive Compilation of Evidence and Explanations about Cold Fusion* (World Scientific), was published in 2007. In May 1993, he was invited to testify before a congressional committee about the cold fusion effect. In 1998, *Wired* magazine honored him as one of twenty-five people who are making a significant contribution to new ideas.

John A. Thomas Jr. began collaborating with John R. R. Searl in 1990 through their mutual friend William Sherwood. In 1993, he wrote the book *Antigravity: The Dream Made Reality*. In 1994, he published Searl's book *The Law of the Squares: Book 1*. He has continued publishing Searl's books on the law of the squares (as of 2010 there have been twenty-four). In 1995, he directed the building of the Searl Effect website (www.searleffect.com) and compiled an information sheet for specifications to build a Searl Effect Generator (SEG). By 2007 he was the holder of a degree: doctor of Searl technology. He is currently working with Searl and Fernando Morris to build the SEG.

Thomas Valone, Ph.D., P.E., is a physicist and licensed professional engineer with over thirty years professional experience, and currently an author, lecturer, and consultant on the subject of future energy developments. He is president and founder of Integrity Research Institute. He has authored six books including *Zero Point Energy: The Fuel of the Future, Harnessing the Wheelwork of Nature: Tesla's Science of Energy, Practical Conversion of Zero-Point Energy,* and *Bioelectromagnetic Healing: A Rationale for Its Use.* He meets regularly with congressional and senate leaders and briefs them on the latest energy developments. His views regarding energy-related matters have been featured on national media including CNN, A&E, the Discovery Channel, and the History Channel. His website is www.IntegrityResearchInstitute.org.

Gerry Vassilatos is the author of *Lost Science* and *Secrets of Cold War Technology* and writes about the remarkable lives, astounding discoveries, and incredible inventions of such famous people as Nikola Tesla, Royal Raymond Rife, T. Townsend Brown, T. Henry Moray, and many others. Vassilatos claims we are living hundreds of years behind our intended level of technology and must recapture this "lost science." His books are fascinating reading for anyone who wants to know about suppressed technologies, which could have prevented many of the global problems we are now witnessing.

INDEX

Page numbers in *italics* represent illustrations.

Abbe, Ernst, 131–32
AC current, Tesla and, 16–18
active consciousness, 77–79
aether, 195, 203
agriculture, 180–81, 214
air pollution, 214
air travel, 214
Alexandersson, Olof, 238
Allan, Sterling D., 169
American Chemical Society, 261
American Physical Society, 261
Amini, Farzan, 194
Anastasovski, P. K., 227
anomaly point, 91–92
antigravity technology, 212
    demonstrations of, 234–36
    Disclosure Project and, 230–33
    evidence of, 229
    historical background, 223–27
    implications of research, 235–37
    recent developments in, 227–29
    suppression of, 156–67
Arata, Yoshiaki, 268
Armstrong, Edwin, 40

Army Ballistic Missile Agency, 224
Arp, Halton, 236
artesian waters, 96–97
ARVs, 230
Aspden, Harold, 238
Astor, John Jacob, 24, 50
Aviation Report, The, 225–26
Aviation Studies International, 225
Aviation Week and Space Technology,
    229
Avouris, Phaedon, 299–300
axial motion, 86–89, *88*

B-2 bomber, 166–67, *167,* 229
*Bad Science: The Short Life and Weird
    Times of Cold Fusion,* 258
Bahnson, Agnew, 163
barium, 268–69
Bearden, Thomas E., 208–9, 228, 239
Beaudette, Charles G., 254
Beck, Christian, 286
Beckhard, Arthur, 16
Behary, Jeff, 67–73
Bell, Alexander Graham, 24

371

Bell Laboratories, 294–95

Bell Telephone Labs, 130

Biberian, Jean-Paul, 261

Biefeld, Paul Alfred, 159–60

Biefeld-Brown effect, 159–60, 224, 229

Binns, Chris, 281, 296

biosphere solutions, 181–85

birthrate, 217

BlackLight Power, 187, 204–7

Blavatsky, Helena Petrovna, 49, 51–52

blood, 80, 110–11

Bloomfield-Moore, Clara, 50–51, 61

Bohr, Niels, 20, 21

Boyce, Bob, 198–202

Braun, Werner von, 224–25

Bremer, L. Paul, III, 310

Bridges, Harry, 133

British Petroleum, 177

Brown, Charles, 285

Brown, T. Townsend, 156–67, *157*, 224, 225, 234

  character assassination of, 163–67

  curiosity of, 159–60

  harassment of, 158–59

  Philadelphia Experiment and, 160–63

Brown, Yull, 200

Brown's gas, 201–4

Brush, Charles, 165

Burridge, Gaston, 157

BX virus, 141–42, 143

C^2, 79, 85, 115, 196

Callen, H. B., 299

cancer, 140–42, 142–46

Capasso, Frederico, 280

Carlyle, Thomas, 177

Carpenter, Scott, 210

cars, 187–88, 190, 200–201, 201–2, 206

Carter, Jimmy, 218

Casimir effect, 227–29, 279, 284, 289–93

Casimir engine, 291–96

Casimir force, 279, 280–82, 289–93, 293, 297, *297*

Catalyst-Induced-Hydrino-Transition (CIHT), 206

cavity quantum electrodynamics (QED), 289–90

Celestine Prophecy, The, 38

CERN (European Organization for Nuclear Research), 258

Channon, Jim, 182

chaos theory, 299

China, 216–17

chlorination, 109–14

Clarke, Arthur C., 223, 254

Claytor, Thomas, 267

climate shock, 172. See also global warming

Close, Frank, 258

coal, 213–14

Coats, Callum, 74–117

cold fusion, 178–80

  benefits of, 265

  disadvantages of, 272–73

  explanation of, 262–63, 270–72

  history of, 257–62

  production of, 265–270, *266, 269*

  value of, 255–57

Cold Fusion: The Scientific Fiasco of the Century, 258

combining frequencies, 26–27

comprehension, dimension of, 76
Compton, Arthur, 20
Congress, U.S., 220
continuous wave wireless, 25–32
Cook, Nick, 233–34
Coolidge, W. D., 148
Coolidge X-ray tubes, 160
cosmic rays, 35–36
Couche, James, 152
Crane, John, 153
Crookes, William, 20
crop circles, 319–24
cryophoros, 20
crystals, 144

Dam, Henry, 52
Dane, Ernest B., Jr., 130
Davidson, John, 238, 239
Davis, Eric, 279
De Aquino, Fran, 234–35
Defense Advanced Research Projects
    Agency (DARPA), 260
De Forest, Lee, 148
Demoyens, Emile, 128
desertification, 214
deuterium, 266–68
deuterons, 272
Dewar, James, 20
dielectrics, 164–65
dielectric value, 92
dimension of comprehension, 76
Dingel, Daniel, 190
diodes, 284–85, 286–87
Dirac, Paul, 283–84
Disclosure, 212–13
Disclosure Project, 230–33
distilled water, 95
DNA, 80–81, *81*

Dodge, Mary Mapes, 24
domination, 217
drinking water, 109–16, 186
Drude model, 294
Dvorak, Anton, 24
dynamic equilibrium, 228
dynamic theory of gravity, 36

Eagle Research, 202
Earth Charter, 325–35
Eckman, Chris, 203
ecotechnology, 84–85
Ecuador, 303–4
Edison, Thomas, 17–18, 22, 23
Einstein, Albert, 20
Eisenhower, Dwight, 160, 210, 212
Electrical World and Engineer, 32–35
electric car, 35
electric entities, 72
electrogravitics, 224–27. *See also*
    Brown, T. Townsend
electrolysis, 188–89, 191–92,
    197–98
electromotive force (EMF), 286
electron microscopes, 142–43
electrons, 228–29, 245–46
Emoto, Masaru, 180
energy, Schauberger and, 80–89
energy-cannon, 107–8, *107*
Energy Research Advisory Panel,
    259–60
energy technologies
    crisis of, 172–75
    demonstrations of, 234–35
    Disclosure Project and, 230–33
    evidence of, 229
    historical background, 223–27
    implications of research, 235–37

recent developments in, 227–29
revolutions of, 178–85
vision of, 236–37
See also antigravity technology; cold
   fusion; free energy; Searl Effect
   Generator; water power
"Energy Unlimited," 278–79
Engelhardt, Tom, 306–14
environmental activism, 325–35
Ernst, Josef, 62
etheric force, 57–58, 195
ETVs, 212, 218, 219, 230–33
Europe, 216–17
Evans, Myron W., 227
evaporation, 114–15
Eversole, Finley, 315–24
Exemplar Zero Initiative, 184, 302–3
extraterrestrials, 211–13, 315–19

Faraday, Michael, 188–89, 197–98
Fate, 157
fecal matter, 82
Feigel, Alexander, 288–89
Feynman, Richard, 280
Fire From Ice, 258–59
First Earth Battalion, 182
First International Conference on
   Cold Fusion (ICCF-1), 261
fish, electricity-generating, 189–90
Fishbein, Morris, 152
Flanagan, Patrick, 180
Fleischmann, Martin, 191, 257, 263,
   264
Fleming, John Ambrose, 20
fluorescent lightbulbs, 41
fly-wheel, 54
Foord, A., 139
four occupations, 306–14

free energy, 175–76, 178–80, 183,
   184–85
   environmental implications, 213–15
   history of, 211–13
   societal implications, 215–17
   support for, 302–305
   vision of, 211, 236–37, 252–53
   world peace and security
      implications, 217–22
   See also Searl Effect Generator
Free Energy Pioneer, 46
Fuller, Buckminster, 175, 179, 182,
   183, 184, 305
fusion, 262–63
Fusion Technology, 260–61

Garbon, Joel, 170, 240
Garrett, Henry, 189, 197
Garrison, Jim, 172
gas, 213–14
General Electric, 22
Generator, 57–58
geopolitics, 216–17
Gernsback, Hugo, 38
Givers, 174
Global Innovation Alliance (GIA),
   183–84
global warming, 172, 213–14, 245, 313
globe motor, 53–54, 59, 60
Goethe, Johann Wolfgang von, 117
Göring, Hermann, 61
Graneau, Peter, 192–94
Graton, Louis C., 130
Gravity Rand Ltd., 225–26
green technology, 175–78
Greer, Steven, 210–22, 234–35, 317
Griggs, James, 195
groundwater, 96

Hammond, John Hays, Jr., 27, 34
Hawton, Margaret, 279
Heaviside, Oliver, 20
heavy water, 270
Heisenberg, Werner, 240
Hertz, Heinreich, 25
High-Frequency Currents, 67
Hinton, Charles Howard, 50
hot fusion, 262–65, 267
Houston, Edwin, 145–46
"How I Control Gravity," 160
Huizenga, John, 258, 260
Hunt for Zero Point, The, 234
hydraulic motor, 54–55
hydro dams, 186
hydroelectric power, 15, 18–19
hydrogen, 197, 200, 279
hydrogen and water chemistries, 178–
    80, 193–94
Hydrosonic Pump, 195
Hydro Vacuo Engine, 54–56
hydroxy boosters, 202–4

Industrial Revolution, 173
Infinite Energy, 192, 194
Interavia, 156
Interstellar Technologies
    Corporation, 292
iron, 269
Ishmael, 173
Italian National Agency for New
    Technologies (ENEA), 261

Jane's Defense Weekly, 234
Japan, 216–17, 262
Jebens, Heinreich, 35
Johnson, Milbank, 139, 149
Johnson, Robert Underwood, 23–24

Johnson noise, 284–85, 287
*Journal of Condensed Matter
    Nuclear Science,* 261
juvenile water, 95

Kanarev, Philipp, 191–92, 195
Kanzius, John, 197
Karabut, Alexander, 267
Keely, J. A., 47
Keely, John Worrell
    discoveries of, 53–60, 189
    introduction to, 43–47
    Kinraide and, 62–67
    life of, 47–52
    pictured, *44–45*
    possible legacies of, 60–67
    Third Reich and, 61–62
Keely motor, 54–55, 71–73
Kendall, A. I., 138
Kennedy, John, 179
Kervran, C. Louis, 270
Khul, Djwhal, 118, 209
King, Moray, 203
Kinraide, Thomas Burton, 62–67, *63*
Kipling, Rudyard, 24
Koch, Roger H., 285–86
Koldamasov, A. I., 195
Komaki, H., 270
Korn, Arthur, 26

Laue, Max von, 127–28
Laurie, Hugh, 38
LaViolette, Paul, 158, 163, 164, 167,
    229
law of the squares, 241–46, *242*
leadership, 217
Le Chatelier, Henri Louis, 271
Leeuwenhoek, Antonie van, 123

lenses, 123–24, 125
Lewis, Nathan, 259
Liberator, 58
Lieber, Charles, 299
Loder, Theodore C., III, 223–37,
    234–35
Lodge, Oliver, 20, 239
longitudinal vortices, 105–9, *107*
low-energy nuclear reactions
    (LENR), 194
Lucas, Francis, 130, 131

Maclay, Jordon, 278–79, 280–81, 297
Maddox, John, 258
magic squares, 241–46, *242*
Mallove, Eugene, 258
Mandela, Nelson, 304
manganese, 270
manganese sulphate, 270
Manning, Jeane, 156–67, 186–207,
    240
Marconi, Guglielmo, 25, 30–31, *30*,
    39–40
Martin, T. C., 20, 23–24
Maxim, Hiram, 50
McCandlish, Mark, 232
McKubre, Michael, 262
Mead, Frank, 284, 300
Merrington, Marguerite, 24
meteoric water, 95
Meyer, Karl, 139
Meyer, Stanley, 197
microdisk lasers, 294–95
microscope, universal, 122–38
Miley, George, 261
Millikan, Robert, 20, 24
Mills, Randell, 204–7
Milton, Richard, 118–19

molecular refrigerators, 288, *288*
Moore, Clarence, 52, 62, 65–66
Moray, T. H., 164
Morgan, J. Pierpont, 22, 29–30, *29*
Morris, Dan, 230
Morrison, Douglas, 258
motion, 85–89, *85*, *88*, 288–89
Muir, John, 24
Muller, Hartmut, 169
Myers, Norman, 181

Naessens, Gaston, 128
NANOCASE, 296–97
nanomachines, 281
NASA, 224, 234, 236
National Missile Defense System,
    218–19
Nature, 76–77, 79, 117
Nature, 258
Navy, U.S., 260
neodymium, 245, 251
New Hydrogen Energy, 262
New Scientist, 228, 279, 280–81
Niagara Falls, 24
Nicholsen, Oliver, 33, 160
Nipher, Francis E., 160
noise pollution, 214
nuclear power plants, 215
nylon 34, 249–50

Oberth, Hermann, 224–25
occupations, four, 306–14
Occupy Wall Street, 308–9
Oechsler, Bob, 229
oil, 173–74, 190, 213–14
O'Leary, Brian, 169–70, 171–85,
    302–5
O'Neill, John, 40

open sourcing, 198–99
order, 298–99
original motion, 85–89, *85*
oxyhydrogen, 202–4

Pacheco, Francisco, 189–90
Paderewski, Ignace, 24
Paijmans, Theo, 43–73
palladium, 266–68, 269
Pantone, Paul, 201–2
Paracelsus, 188
Park, Robert, 258–59
particle beam gun, 15–16, 36–37
Pasteur, Louis, 242
Pauli, Gunter, 174, 176
Pear Harbor demonstration, 224
Pell, Claiborne, 217–18, 220
Penman, H. L., 92
Pentagon, 224, 234
Perkins, John, 176
Philadelphia Experiment, 160–64
Physical Review, 279
physical vacuum, 277
Pinto, Fabrizio, 282, 286, 292–93, 295
Planck, Max, 208, 298
Planck's constant, 278, 297
plasma electrolysis, 191–92
platinum, 269
Poeschl (professor), 17
Pons, Stanley, 191, 257, 263, *264*
Popular Mechanics, 281–82
poverty, 215–17
Prigogine, Ilya, 299
Prismatic Microscope, 135–36,
    138–42
Prodigal Genius, 40
Project Greenglow, 234–35
Project Winterhaven, 162, 163, 224

public utilities, 214–15
Puharich, Andrija, 190
Pukas brothers, 17–18
Pupin, Michael, 24, 39–40
Puthoff, Harold, 227–28, 279

quantum vacuum energy state,
    213–14
Quinn, Daniel, 173

Radford, Arthur W., 161–63
radial motion, 86–89, *88*
radiation, 267
radioactivity, 36
rainwater, 95
Ramsey, John, 260
Rand, Ayn, 36–37
Raven, Peter, 181
RCA, 129–30, 132
Redfield, James, 38
renewable energy sources, Tesla's
    impact, 20–22
Report 13, 166
resonance, 199–201
Rhodes, William, 200
Rich, Ben, 235–36
Rickover, Hyman, 157
Rife, Royal Raymond
    inquisition of, 150–55
    pictured, *121, 135, 155*
    as seeker, 121–22
    universal microscope of, 122–38,
        135
    viruses and, 138–50
Rivaz, Isaac de, 188
Robey, James, 187–88
Robinson, Douglas, 24
robotic boat, 26–27, *27*

Rockefeller Institute, 132
Roosevelt, Teddy, 24
Rosen, Eric von, 50–52, *51*, 61
Rosenow, E. C., 139
Rosin, Carol, 219
Roy, Rustrum, 180
Russia, water science from, 191–92
Rutherford, Ernest, 20, 21

Santilli, Ruggero, 201
Sayano-Shushenskaya explosion, 194
Schauberger, Jorg, 196
Schauberger, Viktor, 114–15, 196
    about, 74–79
    on healthy body and mind, 112–13
    Nature and, 76–77, 79
    nature of energy and, 80–89
    pictured, *75*
    See also water
Schwartzenegger, Arnold, 39
Science and Inventions, 160
Searl, John, 241–48, 252–53
Searl Effect Generator (SEG), 242,
    244–46
    definition of Searl effect, 240–41
    element selection for, 251
    impressed magnetic fields of, 250–51
    structure of, 248–50, *249–50*
    technology of, 246–48, *247*
Sears, Stephan, 207
secrecy, 219
security, 217–22
Seifer, Marc J., 13–42
SETI (Search for Extraterrestrial
    Intelligence), 227
Smoliner, J., 284
SNCASO, 162–63
Sorenson, Brad, 232–33

specific heat, 92–94
Spencer, Herbert, 26
Sperry, Elmer, 24
spider turbine, 194
Spiritualism, 48
springwater, 96
squares, law of, 241–46, *242*
Star Wars, 218–19
Steinmetz, Charles, 22–23, 39–40
Storms, Edmund, 254–75
stress in dielectrics, 158
Stringham, Rodger W., 268
Strong, Frederich Finch, 67–68
Sturgeon, William, 17
surface water, 96
sustainable agriculture, 180–81, 214

Takers, 173–74
"Talking with the Planets," 29
Tapping the Zero-Point Energy, 203
Taubes, Gary, 258
technology solutions
    five revolutions, 177–85
    green innovations, 175–78
    O'Leary on, 171–78
    See also energy technologies
telautomaton, 26–27, *27*
temperature gradient, 97–109,
    103–4
Tesla, Dane, 15–16
Tesla, Nikola, 126
    accolades awarded to, 41–42
    breakthrough with AC current,
        16–18
    deal with Westinghouse, 22–23
    early life in Croatia, 14–16, *15*
    extent of his vision, 32–35
    greater renown of, 23–25

hydroelectric power and, 18–19
impact on the world, 20–22, *21*
later inventions of, 35–39
pictured, *21, 23*
quotes of, 169, 208
resentment against, 39–41
statue dedication for, 13–14, *14*
wireless and, 25–32
Tesla coil, 25
Tesla Roadster, 39
Thales of Miletus, 89
theosophy, 51
thermal conductivity, 92–94
thermal environment, energy from,
178–80
Third Reich, 61–62
Thomson, Elihu, 18, 22–23, 24,
39–40, 145–46
Thomson, J. J., 20
Tiller, William, 170, 180
Timkin, Henry H., 133
titanium, 269
Todd, John, 176
Too Hot to Handle, 258
Transdimensional Technologies,
234
transmutation, 268–69
treasure, 67–73
triboluminescence, 108–9
tritium, 267
Turitz, Gene, 312
"turquoise revolution," 174
Tutt, Keith, 170
Twain, Mark, 24, *25*

UFOs, 211–13, 224–25, 230–33.
See also ETVs
Uhouse, Bill, 231, 232

ultraviolet light, 129–31, 134–35
United Nations, 179, 218
United States, 216–17
universe, 85

vacuum fluctuations, 278–79, 289
vacuum lamps, 126–27
vacuum-state energy, 227–29
Valone, Thomas, 164, 276–300
Van den Broeck, Chris, 288
Vapor Condensation Distillation
unit, 207
Vassilatos, Gerry, 120–55
virtual particles, 289
viruses, 124, 138–50
visionaries, 74–75
Vogel, Marcel, 180
Von Braun, Werner, 219
vortex power, 196
vortical energies, 80–85, *81–82*
Vysdotskii, Vladimir, 196, 270

Wallace, Henry, 189–90
Wardenclyffe tower, 32–33, *34*
water
drinking water supply, 109–16
factors related to health of, 97–109
as living substance, 89–97, *91*
longitudinal vortices and, 105–9, *107*
open sourcing of, 198–99
properties of, 91–94
storage of, 114–16, *116*
temperature gradient and, 97–105,
103–4
types of, 94–97
water arcs, 192–93
*Water Car: How to Turn Water into
Fuel,* 187–88

water power
  age of, 187–90
  BlackLight Power, 204–7
  future of, 207
  harassment of inventors, 197,
    201–2
  more heroes of, 197–98
  overview of, 186–87
  possible sources of, 192–96
  resonance and, 199–201
  water science from Russia,
    191–92
water solutions, 180–81
Wellaston, W. H., 20
Westinghouse, George, 18–19,
    22–23, 39–40
White, Stanford, 24, 29
Williams, John, 232
Willing, Ava, 24
Wilson, Bennet C., 49
Wilson, E. O., 181
wireless, 25–32, 28
Wiseman, George, 198

Wizard, 40
world peace, 217–22

X rays, 129–30

Yale, Arthur, 152
Yurth, David, 184

Zeiss scopes, 140
Zeitlin, Gerald, 68–69
zero point energy, 178–80, 203,
    227–29, 230–31, 234–36
  Casimir engine, 291–96
  conversion of, 281–84, 283, 289–91
  diodes and, 284–87
  introduction to, 276–79
  limitations on, 280–81, 281
  order out of chaos, 298–99
  patents of, 285–87
zero point field (ZPF), 278
Zinsser kineto-baric field propulsion,
    234–35
Zworykn, Vladimir, 129